# The Beekeepers Annual

THE BEEKEEPERS ANNUAL
IS PUBLISHED BY
NORTHERN BEE BOOKS
MYTHOLMROYD,
WEST YORKSHIRE
ISBN 9978-1-912271-51-1
MMXX


SET IN HELVETICA LT BY DM Design and Print
Cover: Last flight of the day. (John Phipps)




EDITOR, JOHN PHIPPS
NEOCHORI, 24022 AGIOS NIKOLAOS,
MESSINIAS, GREECE
EMAIL manifest@runbox.com

# The
# Beekeepers
## Annual 2020

NB

# CONTENTS

# FOREWORD

*John Phipps*

**September 2019**

## *2019*

What a year - and it is not even over yet! Ever-increasing temperatures worldwide, wild fires blazing out of control over the greater parts of South America, Africa and Siberia, ice caps and glaciers melting at a rapid rate, and storms becoming more violent accompanied by wind strengths not previously recorded and resulting in widespread flooding and destruction. And yet there are still those, unbelievably, who reject the reality of global warming.

Many swarms and casts - but not all were as easy as this one to collect! (J Phipps)

And, for our bees so far this year, the death of huge percentages of colonies, particularly in the USA, Brazil and the south of Russia. Whilst many of these colonies never made it through the winter months, others continued to perish at an unacceptable rate due to the uncontrolled use of agrochemicals. Whilst honey yields globally, as usual, varied enormously, the spring and early summer - especially in Western Europe - brought out swarms and casts far above the usual norm. Then, too, the Asian Hornet, *Vespa velutina,* has made more incursions into the UK, and the horrendous small hive beetle, *Aethina tumida*, has been found once again in parts of Italy.

Whilst all the problems outlined above show that greater care and vigilance is required by beekeepers in order for their stocks to survive, there is, surprising to many, the irrefutable fact that 'let alone' or minimally-managed colonies, mostly kept in

3

natural 'bee friendly' hives, or existing in tree hollows in the wild, are not only thriving, but becoming resistant to pests and diseases, particularly varroa. By allowing colonies to swarm, the varroa load is shed by the bees that occupy new sites, and letting nature choose the new queens to take over as head of colonies has given the bees the chance to restore their lost vigour and thus improve their immune systems. This allows them to overcome problems that still beset those bees in 'over managed' hives which are trailed from place to place for pollination contracts or in search of nectar sources, mostly mono-floral, whereas, like ourselves, bees need a varied diet to stay healthy.

As temperatures increase in the UK, it might now be possible to grow more widely plants which are less tender in the UK. Some example of these are featured in the diary/calendar section of the Annual.

## Kim Flottum

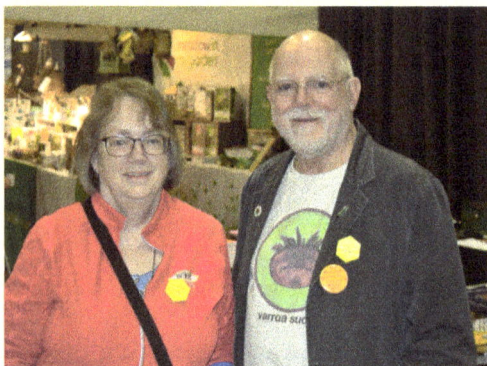

Kim Flottum and Cathy, National Honey Show London, 2017. (J Phipps)

I was sorry to hear that Kim Flottum, Editor of Bee Culture, will be leaving his desk at the end of the year having run the journal since 1986. Over those thirty-three years Kim has been an important voice in American beekeeping and he has shown that of all things he has stood up for bees and beekeepers at a time when the problems for the craft and industry have never been greater. He has given his own unique style to the journal, not only very informative and covering all aspects of beekeeping, but one that is both friendly and homely, reader-centred, thus allowing the scientific and technical items to be easily understood.

Whilst I wish Kim a happy retirement away from his day job, readers will be pleased to learn that by doing so, Kim will have plenty of opportunity to follow up on long-standing projects which work has given him no time to do so, therefore we are likely to hear more from him in the future.

Over many years, I have been very pleased to include many articles by Kim in The Beekeepers Quarterly. How he managed this extra work with twelve deadlines a year and his many visits around the USA, shows what a dedicated beekeeper he is, wishing to share his knowledge of beekeeping as widely as possible.

As a tribute to Kim, I am including two of the articles that he has written for the BKQ as a taster for those who are not familiar with his writing. I say taster, for in the near future we hope to publish all Kim's articles in a single volume, articles which are not only still applicable today, but of interest to beekeepers wherever they may live.

Thanks, Kim!

4

## The Top Bar Hive

Most beekeepers are aware of the top bar hive which allows bees to be kept simply, cheaply and in a bee-friendly way. The hives are easy to make, to manage should you wish to do so, or to let the bees progress unhindered with little or no interference so that the colony can build its comb naturally with the requisite proportion of drone cells, so vital for the well-being of the colony.

It is difficult to put a definite date to when top bar hives became commonly used in the UK. There were certainly long hives, but I became more aware of them when Ron Brown wrote about them in BKQ No 44 (Winter, 1996/7). It is true that there was another hive without frames - Bill Bielby's famous Catenary Hive of which he gave great details in his book Home Honey Production (EP Publishing 1987), a volume devoted to an inexpensive way of keeping bees. Similarly, Ron Brown, writing on economical beekeeping, ended his series with top bar hives, based on his experience on keeping bees in them, firstly in Africa, and later in the UK. This article is reproduced in this issue of the Annual.

Top bar hive made by Ilaria Baldi from Italy. NB Kenyan Flag above the entrance! (John Phipps)

Long hives have been popular for a long time. However, I do not endorse the use of plastics in beekeeping as in this hive produced by Omlet. Bees should live in homes made of natural materials, not those that can become an environmental problem.

## How Will You Celebrate World Bee Day?

On the 20th December, 2017, the United Nations General Assembly adopted by consensus a resolution declaring 20 May as World Bee Day. The purpose of this is to bring to the attention of the public the great importance of preserving bees and other pollinators. It is intended to be a day of action across the world to both educate and to get the public involved in practical projects that will help these insects survive in what is for them an increasingly dangerous world. Whilst the resolution was co-sponsored by 115 UN Member States, including the USA, Canada, China, the Russian Federation, India, Brazil, Argentina, Australia and all the European Union Member States, the original initiative for International Bee Day came from the Slovenia Beekeeper's Association involving three years of negotiations and backed by the country's government.

Two Euro coin to celebrate World Bee Day.

Russian First Day Cover of stamps to celebrate World bee Day.

Hopefully, many beekeeping associations across the UK will make the most of the opportunity that this day provides to inform the public in their localities of the importance of bees and how they can make their areas more attractive for them as well, as warning them of the dangers of using gardening chemicals.

*John Phipps*
*August 2019*

Kim Flottum.

# FROM U.S. TO YOU

**Articles by Kim Flottum written for the BKQ...**

**Unintended Consequences
(BKQ 101, September, 2010)**

In February of 2008 I wrote on these pages about the chemicals our bees were encountering...both inside the hive and out in the real world of foraging for and finding food. I was optimistic then that perhaps beekeepers and farmers were changing directions and that the world would soon be a safer place for us, for our children and for our bees. I was encouraged because when the light was turned on, we could see that we were killing ourselves and our future with the chemicals of agriculture. And the need for change was clear. I was wrong.

Let me tell you a story about unintended consequences. Perhaps this story is one of those Urban Myths you often hear of... plausible but not completely accurate... because I don't have the references to bear it out. But read on and see if it doesn't sound right...

A recent study, I think it was in Florida, was undertaken because there was an unprecedented loss of frogs from small and large ponds across much of the state. The scientists who began the study suspected all manner of things...homeowner pesticide runoff; global warming; a new, exotic disease, human pharmaceutical contamination in the water...anything and everything was up for grabs. What with bees and bats crashing and burning everywhere, who knew what it could be.

What they found wasn't what they were expecting, however. It turns out that the agricultural herbicide atrazine, a chemical that inhibits weeds from growing in crop fields, was washing out of the fields it had been applied to and was ending up in lakes and rivers and ground water and eventually in ponds near and far from where the chemical was applied. These were the same ponds that the frogs were disappearing from. This can't be the problem, they said, because hundreds of tests over decades have shown that atrazine doesn't affect frogs if they had to live in water polluted with small amounts of the stuff (very small amounts of the stuff), and it was only very small amounts they were finding in the ponds. So it couldn't be that, could it?

7

So they continued looking. The next level of tests discovered a very undramatic finding. So undramatic it was almost overlooked. It seems that atrazine, remember it's a plant killer, was once in the pond water, killing the green algae bloom that always grows on top of the water of most ponds in Florida. But even the tiny, tiny amounts of atrazine found in those ponds were enough to kill something as fragile and sensitive as this harmless green algae. Well, it made sense that atrazine could kill algae, even though it wasn't supposed to be there...but how could that harm frogs?

So the scientists brought in some algae specialists. Well, to them it was a pretty simple problem. once the green algae that floats on top of the water is gone, sunlight, a rare commodity in the bottom of ponds with green algae on top is able to filter to the bottom of those ponds, and down there another algae lives, a brown algae. Normally, when sunlight is mostly filtered out by the green algae on the surface the brown algae on the bottom is limited in how much it can grow. It needs sunlight to thrive and limited light means limited growth. When there's lots of light though...watch out.

What happens is this. Brown algae is the primary food source for some of the snails that live in these ponds. Suddenly, lots of food means lots of snails. But these snails have a parasite that lives part of its life in the snail and part of its life in a frog. When this parasite reaches a certain stage it leaves the snail and migrates to a frog. Once infected by this stage of the parasite, the parasite kills the frog in the process of finishing its life cycle. When the frog dies, the next stage of the parasite finishes its cycle, leaves the corpse and goes back to a snail and the cycle begins again. I think I have this right...but I'm neither an amphibian or crustacean parisitologist so I may have this not exactly correct.

Nevertheless, frogs were dying by the score because of the Unintended Consequences of a herbicide application made long ago and miles away. I mention this example of Unintended Consequences because it is similar to another Unintended Consequence that is occurring in the food producing fields of the world that is having a direct and deadly affect on the bees I keep and the food you eat.

Plants, like animals, are plagued by all manner of pests – think frogs and snails and parasites. Bacteria, fungi, viruses, pests and predators all have some role...everything is something else's lunch. For food crops, if the bad guys win there's no food, or at least a lot less of it. So farmers strive to protect their crops anyway they can. And, as you would imagine, as inexpensively as they can... food production is a very, very thin margin business.

Lots of crops are attacked by fungi and bacteria. These organisms aren't all that aggressive so they need to find a plant's most vulnerable spot...for some plants it is the stomata, where water vapor goes in and out of a leaf that is vulnerable, but for other plants that sweet spot is the flower. Flowers are fragile and susceptible to damage from the weather, from animals eating them, and from invaders at the microscopic level. And an infected flower usually mean the flower dies so no fruit is produced, or the infection lingers and the eventual fruit becomes infected later. In either case Farmers have to protect the flowers...and to do that they have to apply pesticide sprays directly on the flowers themselves. Sprays that attack and kill the invaders when they are most exposed and most affected by the chemical applied. Many of these sprays are actually absorbed by the flowers to protect them all during bloom.

Spraying with agrochemicals has unintended consequences, often unthought of, nor cared for, by the those operators who carry out the work. (J Phipps)

Here's where the Unintended Consequences come in. It seems that the new generation of fungicides that farmers are using are very effective at handling the fungi and bacteria that are invading the flowers. But they are also very effective at killing the pollen grains that are trying to fertilize the flowers being sprayed. That's not good. The protective agent is actually killing the host it's supposed to protect. Pollen grains, as you probably know, actually land on the receptive part of the flower and grow a tube into the flower tissue so the male part of the pollen cell can reach the female part of the flower, and fruits and nuts and vegetables happen. It's that pollen tube stage that's susceptible to the fungicide, it seems.

Pollen which lands on the stigma of the flower grows a tube. It is at this stage that the pollen is susceptible to fungicides, thus preventing fertilisation from taking place. (J Phipps)

But there's more. The pollen that my bees collect from these blossoms has the residue of these chemicals on and in it, too. Bees collect the pollen as a protein source in their diet, and in the process spread pollen amongst the flowers they are visiting and the flowers get pollinated...well, they did until this stuff was sprayed on them. But the collected pollen is contaminated with this stuff too, and bees eat it. And then...

And then some really nasty things begin to happen. The chemicals on the pollen are ingested by the foraging bees and effectively kill all the enzymes and micro-organisms in the bee's gut that help it convert nectar to honey and digest all the food bees eat. That's what these chemicals were meant to do after all... kill micro organisms. The result is just similar to when you take an antibiotic - you have digestion problems because the antibiotics kill indiscriminately...both the good and bad bugs in your system...and so it is with the bees. In fact, when bees

9

come into contact with this stuff they can't digest food and they can't make honey. And you know that bees share food, nectar and feed young, and when they do, they share these chemicals too, so soon nobody can make food or honey. And before they know what's happening they have brought this stuff home and stored it for later use. A chemical time bomb, just waiting to kill again.

Here's what happens next. Pollen needs to ferment, just like silage for cows, before bees can digest it...and that's what those little guys are supposed to do. Once fermented, the pollen can be digested by the bees and used to make worker, drone and royal jelly. But unfermented pollen can't be digested and it can't be turned into jelly and guess what...bees can starve to death trying to eat more and more and more food that they can't digest. There's an Unintended Consequence for you.

Top left: Pollen is eaten by the bees to provide larvae with their food from the hypopharyngial glands. Contaminated pollen can lead to the deaths of both bees and brood.

Above: Pollen needs to ferment in the comb for the bees to get the most benefit from it.

Left. Dead bees being collected from the front of theses once strong hives in Ukraine. A nearby rape field had just been sprayed. (J Phipps)

For farmers, it's damned if they spray, and damned if they don't spray. For beekeepers, they're moving out of the killing fields, and for people who like to eat... well, try some wheat today, or corn...maybe some rice.

# Using Eight Frame Hives

**(BKQ 90, December 2007)**

**Kim Flottum**

I'm getting old. The American Association Of Retired Persons (AARP) sent me a membership card five years ago. Social Security sends me notices on a monthly basis about my status as a soon-to-be old person. I reach normal retirement age, now 66, in six years. That's when I can officially, finally, retire and collect my government pension and begin receiving my company retirement checks.

But I still have these bees in the back yard. Bees that I have, and want, to take care of. Lifting. Moving. Supering. Harvesting. Lifting. Carrying. Lifting. Do you see my point?

Ten-frame equipment is heavy. Deep supers (9-1/2" tall) are heavier than mediums (6-5/8" tall), and mediums are heavier than shallows (5-11/16" tall) – but they all are heavy when they are full of honey, bees and wax.

There's a solution. I've switched to using equipment with eight frames instead of 10 frames. They are 20% lighter no matter what size you use… deep, medium or shallow supers. But an eight frame deep super, when it's full, still weighs about 60 pounds (that's like 27 kilos or so I think), and a day of lifting even these is more than my nearly retired back wants to endure.

I don't think any bee supply company makes eight frame shallow supers, but eight frame mediums are pretty common. And that's what I use, and that's what I like. An eight frame medium super, when it's at its heaviest full of honey only weighs about 40 pounds...18 kilos or so...So I'm still keeping bees and I'm still having fun, and my back doesn't bother me when I'm done working my bees. That's a good thing.

But you keep bees a bit differently when you use all mediums and all eight frame equipment. It's not difficult, but you should be aware of the differences so you don't make the same mistakes I've made over the years. We'd appreciate it if you made new mistakes and then shared them with us. That way we'll cover more ground.

For starters, common sense says that you will simply need more boxes for each hive and more frames to fill those boxes. I dealt with that issue easily because I purchase pre-assembled equipment. The supers are already glued and nailed, and the frames come glued and nailed with a sheet of plastic foundation already in place. I simply take a super, with the frames already assembled and already in place, out of the shipping box and put them on the hive. I let the supers sit outside for a season, sometimes two, before I add paint or stain since I'm always late getting them ordered and just never seem to have the time before I need them. Wood preservation is not at the top of my list of important things to worry about.

Wait. Pre-assembled equipment? Absolutely! I will never, ever put together another frame in my life again. Never. When you stop to consider the time and energy assembling beekeeping equipment takes, it doesn't make a lick of sense. An unassembled, unpainted medium super costs about $10.00, plus paint, nails, glue and time. Unassembled frames with plastic foundation cost about $1.75, plus glue, nails and time. A full super then goes for about $25.00 plus your labor. A completely assembled, unpainted super with assembled frames goes for about $35.00 and essentially no labor. Figure it takes, what, at least a half hour, probably more to assemble and paint a super, and then paint it again… is your time worth that? Mine isn't. Time is the thing I have least of in life. Remember, I'm getting old. I don't have much time left. Nailing frames and fingers isn't the way I want to spend it.

But back to beekeeping. I use three medium supers as my brood space. Normally, at least here in this part of the US (northern Ohio, near Cleveland), beekeepers use two deep hive bodies for brood space. This is because they need about 70 pounds of honey stored in these two boxes for overwintering. That much honey, in deep hive bodies will take about 12 frames, more or less full.

If you calculate the total available space inside two deep hive bodies you end up with 20 frames X 9000 cells = 180,000 individual cells (there are roughly 9000 cells on a deep frame at 4500 per side – 50 cells tall, 90 cells wide). Of this, the bees need about 60% for brood rearing, or approximately 100,000 cells. That's about 12 frames worth, total, for raising brood. Three medium hive bodies will support 24 frames X 5400 cells = 129,600 individual cells, or about 30% less total space (there are roughly 5400 cells on a medium frames at 2700 per side – 30 cells tall, 90 cells wide). That means the bees, to raise the same amount of brood need about 20 of the 24 available medium frames. Having less space for storing honey isn't a problem but it must be taken into consideration when practising swarm management in the summer and preparing enough honey stores for the winter.

There are times when I will add a fourth medium brood box to my stack of three when I have a rapidly building colony that needs additional room. When I do this I am essentially preparing the colony to be split. When the colony is ready and I have a new queen in hand I will use the newly added brood box as the base of the new colony I am making. To make the split I'll add about eight frames worth of brood spread out over 10 frames that I'll put into two eight-frame boxes – six frames worth of capped brood and about two frames worth of open brood, and enough attending bees to take care of them all. That new queen will be added in a push-in cage over sealed brood and my split will be ready to roll in about a week or ten days with lots of new bees, still some capped brood, a new queen and a lot of room for all the eggs that new queen will produce right away.

Honey supers, too, are more numerous when using eight frame equipment. I don't produce a monstrous honey crop most years because I'm not in a great honey-producing area. But bees will be bees and there's always some extra honey to handle. Moreover, I try to make sure there are at least two full supers of honey for winter feed so I always want some honey production.

An average honey crop in this part of Ohio, most years, is about 70 pounds... sometimes more, often less. A medium frame of honey when fully capped weighs roughly five pounds. It may be as much as six pounds if the cells are well drawn out and full. So a medium super with eight fully-drawn and capped frames of honey weighs about 40 pounds. Seldom more. An average crop just about fills two medium supers. Since the bees almost never fill every frame on both sides wall to wall with honey, figure three supers, each about two-thirds full.

But wait. Recall that that 70 pound crop doesn't include the honey normally stored in the brood nest area when you're using 10 frame equipment... and that's an additional 40 and maybe as much as 60 pounds. That means at least one more medium super must be added just on top of the existing three or four brood boxes just for brood food and overwintering. Remember, there's not much extra room for storing honey in those three medium supers used for brood rearing. They are almost all brood. Four medium hive bodies then is probably the best long-term arrangement for brood and stored honey.

So how high it this stack going to get? Let's see. There's at least four, and for awhile five medium hive bodies, and three, maybe four honey supers... that's, what, maybe nine supers high. That's a stack six feet tall, not counting that fancy copper roof and the two-inch deep bottom board.

I'm just exactly six feet tall. So my colony and I are equal. Well, not quite. The hive stand I use - that's a good 18" tall. It's that tall for two good reasons. First, when I take a super off the top of that stack, no matter how tall the stack is, without a decent hive stand that super goes all the way to the ground. Then, when I have to put it back, I have to go all the way to the ground, lift and grunt and groan and get it all the way to the top again. By having a stand like I do the lifting is much, much reduced. This is why I went to eight frame equipment in the first place... no more heavy lifting.

The second reason is skunks. Colonies that sit just off the ground are prone to foraging skunks. When a skunk, and usually her three or four kits forage at the front door of a colony they slap and scratch the landing board until guards come out to investigate. The skunk slaps a guard bee, grabs it and eats it. Sometimes the bee is only stunned and it will sting the mouth, tongue, throat or even the stomach of the skunk. This apparently doesn't bother her and she will sit and feast until she's full. A skunk family can effectively kill a colony in a two week period if something isn't done to stop the foraging. Some beekeepers go to extreme lengths to stop them, using boards with nails in front of the colony or those tack-like carpet grabbers on the landing board. Rolls of wire fencing in front of the colony are common too, so the skunk can't get near enough to eat. All of these are dangerous and extra work. Because of the height of my hive stand skunks can't reach the landing board so that hive stand saves my back, my feet and my bees.

But the best way to keep my stack of boxes at a reasonable height is to extract more than one time in the summer. I try to get the light, delicate honey in the spring, the more flavorful summer honey and the robust fall crop all in their time, so that I actually have three kinds of honey to sell or give away. You'll always sell more if you have three kinds of honey rather than a single, medium colored, no-distinct-flavored honey to sell.

Eight frame equipment fits my style of keeping bees. The boxes are easy to handle, even in the middle of summer the height of the stack is kept to a minimum by timely harvests, and management is similar in principle to 10 frame management but simply requires a bit more attention to detail. All things considered, size eight fits me a lot better than size 10.

If you are getting old, if your back is telling you that beekeeping may be too hard, if your knees complain about carrying heavy honey supers in the heat of the summer I suggest you consider switching to eight frame equipment. Your body, your bees, and your spouse will appreciate the change. Try!

My apiary amongst which are 8-frame hives.

# APIARY WORK

**R Raff, BKQ 70**

John Gleed, alias R. Raff, Nairn, Scotland. (Photo: J Phipps)

John Gleed with his 'Coffin Hive'. (Photo: J Phipps)

It is nice to go into a well-kept apiary and it gives an indication as to the kind of beekeeper who works there. Not long ago I was involved in doing a tidy up job in an apiary where the weeds and grass were almost waist high.

There were two others there besides myself, and they arrived with petrol strimmers and there was also a third strimmer as a back-up.

I had my scythe with the blade wrapped in a piece of blanket and tied with rope and a whetstone sticking out of my pocket. You can't get much more basic than a scythe and I felt a little embarrassed in the face of all the heavy metal the other two had lined up for the job. The plan was that I would take down the rough, rake it up, and then they would come behind with the strimmers making a nice tidy job of it and with that arranged I got started. One of the other two was having a deal of trouble with his strimmer and no amount of yanking at the cord produced more than a few half-hearted coughs. Eventually he said he would have to take it away into town and get someone to look at it. Before doing this he tried up the spare strimmer but it didn't even give a cough so he gave up and set off for the town. I kept mowing and raking up and my mate continued with his strimmer but he said he was shortly going to have petrol trouble. Half of the petrol in the little can he had brought had gone in the strimmer that was now away to the town. However, he kept going and between us we were making an impact. Then the man returned from town but minus his strimmer. He said he would take over the strimming and the man with the strimmer said he would do the scything which left me just pottering about raking up.

This being so, I had ample time to take stock and especially to study the technique of the man now on the scythe. I'll say this for him, he certainly had style because he set about those weeds with the vigour of a whirling dervish. With arms fully outstretched before him he would raise the scythe away to the right above his head and then with a tremendous swing it would come crashing down. I watched for a while and wondered maybe if I should have a word with him. I had whetted the blade till it was like a razor and now I was worried in case he slipped or tripped and ended up slicing off his napper, but at the same time I did not want to embarrass him. He told me previously that he came from North London, Wembley, I think he said, and I supposed they probably did not do a lot of scything around there. I decided to keep quiet and hope for the best because he was slashing the stuff down after a fashion. Meanwhile the strimmer man was working away but said the petrol was on the point of running out and very soon after that the strimmer stopped.

This left me feeling exceedingly chuffed about my scythe. Here it was still going strong whereas the hi-tech stuff had all ground to a halt. I should have kept these thoughts to myself and taken heart of the Good Book where it tells us "Pride goeth before destruction, and a haughty spirit before a fall." Destruction for my poor old scythe was just around the corner. Suddenly I heard the most almighty crack and this proved to be the point of the blade hitting a four inch concrete block that was buried in the grass and weeds. It had hit it dead square on and when I went over to see what was what, the mower was examining the upturned scythe which had the blade hanging down and waggling about like a seagull with a broken wing. Five rivets securing the blade to the back rib had popped and the rib itself was badly bent and buckled so now that was four implements out of action. It was fortuitous that when this happened the job was practically finished. There was a day when the village blacksmith would have repaired this damage in half an hour or so and just charged a few bob, but I will probably have to do it myself. It's about as bad as trying to get a typewriter repaired. Mechanics with this skill are as scarce as hen's teeth and charge the earth.

This job we did was just a cosmetic exercise and it would not have mattered two hoots to the bees as long as the grass was kept down in front of the hives. There are other practical things though that can be done to help both the bee and the beekeeper and are well worth the time spent on them. Winters now are not nearly as severe as they were even 20 - 30 years ago. In past years the snowfall was heavier and it lay longer. Bees were confined for much longer periods and when the temperature rose briefly and the sun came out the bees would take the opportunity to make a cleansing flight. It was on occasions like this that hundreds of bees might be found in the snow in front of the hives all dead or dying. The sun on the snow seemed to confuse them. To prevent this we would clear the snow in front of the hives, a strip maybe 1.5 metres wide all the way along. Once the sun got on to what was left in this strip it soon melted and provided there was no more snow there was no more trouble. Some would scatter ashes or lay boards, anything indeed to break the white expanse. It was always a sad sight to have all these poor bees lying there. I used to try picking some up and warming them but it was no good.

Sometimes, if the snowfall was heavy or the snow had drifted, the hives would be covered and there was a temptation to dig them out in case the bees suffocated. Well, there was no fear of this happening even though the entrances were blocked. The snow is porous and to have the hives covered could indeed afford a measure of insulation. The time to act was when the thaw came and then the roofs and entrances were cleared. It is many, many years since I last had to do any of this.

Snow is not a problem to bees as long as they are not drawn out by the bright sunshine. However, sometimes the snow at the hive entrance can turn to ice and deprive the bees from air. Editor's apiary in Lincolnshire.
(Photo: J Phipps)

A windbreak can make a big difference though to the bees comfort and remember, anything you do like this helps them to do their job better. All it needs is a few posts and wire to which can be fastened any material such as old raspberry canes, spruce branches, twigs like those for staking peas, and branches of broom. Just enough material to break the force of the wind and let it filter through. The bees will really thank you for this.

A supply of drinking water in the spring is a great help. Get something like a an old dustbin lid and put some stones in it for the bees to land on. Bees like water with the chill off it so a little shelter rigged up over the lid will do this. A framework covered with polythene is ideal. It also protects the water from the bees as they fly over it.

It may seem odd to add something here about killing bees but sometimes for the good of all this is necessary. Before American Foul Brood was a notifiable disease I frequently killed affected bees and burned them, the frames, the wax and the honey. The hive parts were scorched and used again. AFB was a scourge just after the war, but I haven't encountered a case now for a very long time. When, for good reasons, a stock needs to be killed, our old method could be used. Half a pint of petrol (0.3 litre) will do the job. When cars were basic we used to disconnect the pipe to the carburettor and pipe the petrol into a can but I don't know that I would like to tamper with a modern car. When all the bees are at home, close the entrance tightly with a cloth. Gently remove the roof and inner cover, pour the petrol over the tops of the frames and quickly close the hive. For a few moments nothing happens but then there is a great hiss and roar which only lasts a brief time - and then silence. That's it. The deed is done!

Some time very soon I will have to set about repairing the damage to the scythe. So, if you happen to hear an anvil ringing the vesper chime, you will know that that is me at work.

Top left: John Gleed kept bees in a variety of hives, most of which he made himself.

Above: He cut both straw and reeds to make skeps.

Far left: Skep shelter.

(Left): Bell jar full of honey from the top of a skep.

Ron Brown

# Make A Top Bar Hive

*Ron Brown*
*(BKQ 44 - WINTER 1996/7)*

Here is something completely different; a hive using no frames, no supers, no queen excluder, no sheets of wax foundations, environmentally friendly and costing very little, attractive in any garden. Based on the principle used by the Greeks 2,000 years ago in their basket hives with sloping sides, but forgotten for centuries. They realised that when side walls are inclined to the vertical by 14 degrees or more, the bees do not join their combs to them as they do when the sides are vertical. Economical in use, involving no lifting of heavy boxes, yet with all combs readily available for inspection at all times, whether for checking queen cells, re-queening or harvesting. A similar hive about one third the size of that shown can be used as a nucleus; in fact, a miniature version in polystyrene with three or four tiny combs is in use by the Germans as a queen mating mini-nuc.

## NATURAL MATERIAL

As can be seen from the photographs, the basic materials are the bark-lined strips which sawmills take off a tree trunk to square it up before sawing it into planks. Of little commercial value except to gardeners, who sometimes use this natural produce to make plant containers. The main body of the hive is made from three similar lengths, each 3ft long by 8 to 9" wide, with two end pieces cut into the shape of a trapezium measuring 12" at the top and 8" at the base. Even these measurements are not invariable, as no frames of constant size are involved and the top bars can easily be cut to fit whatever size you choose, as the bees make their own bee space, so long as the sides are inclined at an angle of at least 14 degrees to the vertical. If the sides are vertical, as in an orthodox hive, the bees will attach their combs to them and so these can no longer be lifted out at will for inspection.

The top bars are cut from thin strips of wood 1 3/8" (3.5 cm) wide and fit into a rebate cut into the side pieces. The model illustrated (in use at Cockington Apiary, Torquay) has 25 top bars, each 12" (30.5 cm) long. In order to get the combs built correctly one can provide starter strips of wax 1/2" deep. In Central Africa over 40 years ago, I had no access to wax foundation so used a carpenter's scribe to cut a groove down the centre of each bar and ran melted wax down this groove. I thought I had invented this idea, but years later read of Thomas White Woodbury, of Exeter, doing much the same in 1859!

*Top bar hive showing entrance at one end of the hive*    *Top bar hive with lid removed*    *Removing combs - this is much easier once bees have been raised in the cells*

**Top bar hive showing entrance at one end of the hive. Top bar hive with lid removed.**

Beekeeping is like this and one often 're-invents the wheel'. In practice the bees did sometimes build their combs at an angle across the bars and complicate my work, so today I prefer to stick thin strips of foundation into the grooves. Once you have such a hive, you can start up a second one by using combs of bees from the first, but to get going in May or June, it is easy to run in a swarm. As with any hive having combs the 'warm way' the bees will store honey at the end furthest from the entrance. This is where comb honey (with no brood) can be harvested at any time by cutting out and then replacing the wooden strips, taking care to leave 1/4" of wax in place for next time. Comb renewal is simple, as the oldest combs are always nearest to the entrance and after removal the others can be pushed in to close the gap.

**COMB INSPECTION**
It might be expected that bees would propolise the bars heavily and make it difficult to move them. In practice this does not seem to happen, so long as the bars are pressing close against each other to make a continuous area of wood, like a cover board. A smear of vaseline along the rebate helps to make it easier to push bars along to close up any gaps. With no supporting wire or frames newly built combs are fragile and must be held vertical at all times, never at an angle. Once a comb has been bred in it is much stronger and can be handled much more readily. One soon learns the knack of doing this, as shown in the photograph.

**HONEY HARVEST**
Obviously the natural combs can be cut out and used (or sold) as they are, or chopped up to fit into standard 8 oz plastic containers. It would be very difficult to centrifuge these combs as we normally do. In tropical Africa 40 years ago, I used home-made hives, but with vertical sides, these were so much harder to work. At no time did I have an extractor, yet regularly bottled and sold large quantities of liquid honey at 3/6d a pound, having no competition except from imported honey. This price (17p today) may sound very modest, but honey in the UK was only 2/6d a pound at that time. Of course I also sold honey combs by weight, preferred by many customers, including one or two living in Devon today but they have to pay much more now!

## FEEDING

In Africa I never needed to feed bees and even in the rainy season they were able to maintain themselves (but of course, gave no surplus). Here in Britain one can easily feed at the rear of the hive, using bakers' fondant. An ingenious beekeeper can make or modify a feeder for use with syrup if desired.

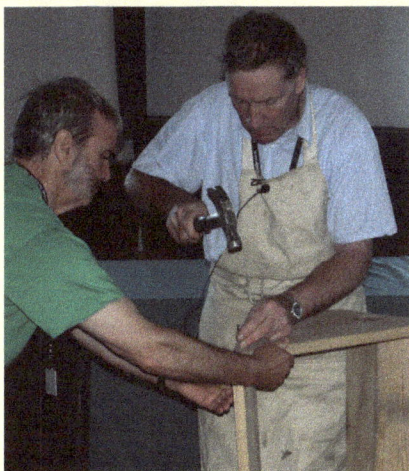

A top bar hive can be made with just a few tools - a saw, a hammer and a plane. The hive here was created in one hour during the Apimondia Congress in Ireland in 2005, except for the top bars.

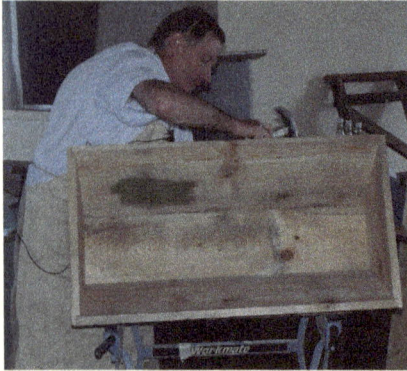

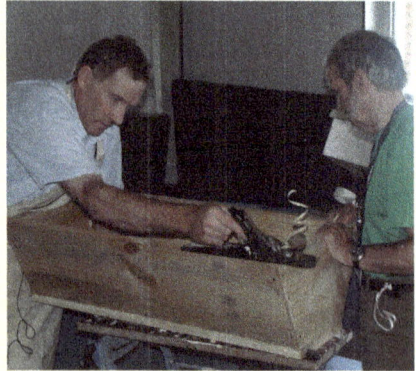

## ROOF

The hive in use at Cockington (made by Roger Kirk to my specifications) has a roof made of thick plywood framed by wooden bars, having a waterproof felt top. Because of its large area it is securely tied on for winter.

## Sir Edmund Hillary, KG ONZ KBE

*20th July 1919 - 11th January 2008*
*Beekeeper, Explorer, Climber and Philanthropist*
Sir Edmund Hillary

Sadly, today, Mount Everest has been spoilt by an ever-increasing number of mountaineers, mostly wearing their high tech clothing and with their equipment contained in a single back-pack, hoping to delete from their bucket list their conquering of the world's highest mountain. Not surprisingly, many have been defeated by Everest leading to search and rescue teams having to risk their own lives to retrieve the injured or dead. With so many attempting the climb there are often queues to reach the summit, and horrendously, the amount of litter left behind by the climbers has despoiled what was once one of the world's most remote and sacred of places for the ethnic population.

How different it was all those years ago in 1953, when Hillary and Sherpa Tenzing, members of John Hunt's expedition, reached the peak on the 29th May, the first ever men to achieve this momentous feat. The news reached the UK at the time of Queen Elizabeth's coronation, giving its people a double cause for celebration.

I was only five years old at the time, but still remember the crackled radio reports from such a long distance away. The following year, beginning at last to read reasonably well, I took every opportunity I could to dip into John Hunt's The Ascent of Everest, published in huge numbers for members of the Companion Book Club, mostly to wonder in awe at the exciting pictures and to learn the names of so many members of the team. I still have a copy of the book.

When I started beekeeping, learning that Edmund Hillary had come from a beekeeping family re-kindled my interest in his life and career. The family ran 1600 hives, the heavy work being carried out by his father Percival (who was also the editor of the New Zealand Honeybee), plus Hillary and his brother. His mother Gertrude also contributed to the business, specialising in breeding and selling queen bees. Withe less beekeeping to do during the winter months, Hillary was able to spend his time climbing, a passion which eventually took him to the top of the world.

NZ bank note in honour of Sir Edmund Hillary.

On behalf of the New Zealand Beekeeper's Association, Frank Lindsay wrote in The New Zealand Beekeeper:

*"After Everest he only did one more season with the bees. His real legacy is in the Himalayan trust he established to help the Sherpa people of Nepal. Together with trusts he helped set up in many other countries, he raised the money for schools, hospitals, a runway, water systems, bridges and many small village projects but the remarkable thing was that with Rex and climbing friends and associates he would go over there and help complete the projects with the Sherpas. Even when he was New Zealand's High Commissioner to India, he took time off to visit the Sherpas and was seen hammering in nails to fix a leaky school roof. His son Peter said at his funeral that he was perhaps the only person in the world that would take a prefabricated building on an aeroplane as baggage. Peter said school holidays were looked on with excitement and perhaps a little dread. They never knew where they were going but he said it was the most memorable thing "having adventures with your father". It was not unusual to set off with two rafts, strapped on to a Mini Cooper (one on the roof and one on the back). The whole family of six climbed, hiked, camped, rafted and had holidays together.*

*Peter and Sir Ed went to the North Pole and the next year Sir Ed was the first to overland to the South Pole using Massey Ferguson tractors. He was meant to only set up supply bases for Sir Vivian Faulks but after waiting a few days at the last dump, he calculated that they had enough fuel to get to the Pole. He got there with fuel for only four kilometres left, a few weeks ahead of Sir Vivian. His punishment was that he had to sit in the back of a snow cat (without windows) all the way back to Scott Base, only to be let out to give navigation instructions whenever they got lost.*

*In 1977 he travelled up the Ganges in jet boats where the banks were lined with millions of people waiting to see him. On a climb at the end of the trip, they went to 18,000 feet too quickly and Sir Ed got altitude sickness and had to be carried down to a lower level where he revived. This stopped him going to high altitudes again.*

*Sir Ed went back to the Himalayas and to Scott Base to celebrate their 50th anniversaries with the help of a doctor and oxygen.*

*What set Sir Ed apart was that he was down to earth and always approachable. He was also a modest man living in the same house he, Rex and his father built in the 1950's. All the money he raised went to others. He was a good team leader and a meticulous planner producing sheets of paper as to what each member of the team would be doing each day. But he was also adaptable; when a plan went wrong he would change it to achieve the goal.*

*The National Beekeepers Assn of NZ presented the family with a posy of wild flowers in a smoker. We are very proud to claim this man as a great New Zealander and a beekeeper. We mourn his passing but reflect with admiration on what he achieved in his 88 years"*

## Quotes by Sir Edmund Hillary:

1. No one remembers who climbed Mount Everest the second time.
2. You don't have to be intellectually bright to be a competent leader.
3. As a youngster I was a great dreamer, reading many books of adventure and walking lonely miles with my head in the clouds.
4. You don't have to be a hero to accomplish great things—to compete. You can just be an ordinary chap, sufficiently motivated to reach challenging goals.
5. People do not decide to become extraordinary. They decide to accomplish extraordinary things.
6. I have been seriously afraid at times but have used my fear as a stimulating factor rather than allowing it to paralyse me. My abilities have not been outstanding, but I have had sufficient strength and determination to meet my challenges and have usually managed to succeed with them.
7. Life's a bit like mountaineering – never look down.
8. Mount Everest, you beat me the first time, but I'll beat you the next time because you've grown all you are going to grow...but I'm still growing.
9. While on top of Everest, I looked across the valley towards the great peak Makalu and mentally worked out a route about how it could be climbed. It showed me that even though I was standing on top of the world, it wasn't the end of everything. I was still looking beyond to other interesting challenges.
10. It is not the mountain we conquer, but ourselves.
11. I have never regarded myself as a hero, but Tenzing undoubtedly was.
12. If you have plenty – more than enough – and someone else has nothing, then you should do something about it.
13. With practice and focus, you can extend yourself far more than you ever believed possible.
14. If the going is tough and the pressure is on, if the reserves of strength have been drained and the summit is still not in sight, then the quality to seek in the person is neither great strength nor quickness of hand, but rather a resolute mind firmly set on its purpose that refuses to let its body slack or rest.
15. Motivation is the single most important factor in any sort of success.

16. I have discovered that even the mediocre can have adventures and even the fearful can achieve.
17. The explorers of the past were great men and we should honour them. But let us not forget that their spirit lives on. It is still not hard to find a man who will adventure for the sake of a dream or one who will search, for the pleasure of searching, not for what he may find.

**EVEREST FACTS**
- Height: 29,029 feet (8,840 meters).
- Location: Himalayas - Nepal and Tibet part of China, 160.51 km from Kathmandu.
- Names: Everest after discoverer Sir George Everest in 1865.
- Nepalese - Sagarmatha (Forehead of the Sky).
- Tibetan - Chomolungma (Goddess Mother of the Mountains).
- Members of British Expedition, Sir George Mallory and Andrew Irvine might have reached the summit in 1924, before "vanishing in the clouds."
- Edmund Hillary and Sherpa Tenzing reached the summit at 11.30 am, 29th May, 1953. They buried sweets and a cross in the snow before they began their descent.
- Despite there being at least one death on Everest nearly every year, it is supposed to be an easy mountain to climb except for altitude problems. In 2016, 641 climbers reached the summit with only the deaths of 5 climbers.The most dangerous high mountains are considered to be Annapurna and K2.
- Commercial exploitation of Everest has made it the highest rubbish dump in the world. Now there are strict laws compelling climbers to bring down their rubbish or face heavy fines.

## Honey as a Business

*John Phipps*

The huge yellow fields of oil rape which continued to increase in number every year almost guaranteed a good harvest of honey every season and, apart from one or two exceptions due to poor spring weather, I normally had plenty of honey to sell. My workplace provided an important outlet, sales from my gate were usually good, and one or two local shops and a garden centre would each take a couple of dozen jars at a time. Selling wholesale to shops nearby was not easy though, as many of them had procured their honey from the same beekeeper for many years, so it was really a case of waiting for dead men's' shoes, if I was to be their main supplier. In the few places that I managed to sell my honey I often wondered what service I got for the 30% profit the shop owner received. There was no real attempt to promote my honey and it was sad to see the jar lids gathering dust in a corner of a gloomy shop. I often mentioned making some sort of display, but no interest was shown; after all, my honey was just another of the hundreds of commodities that the shop had for sale.

I was really envious of Jeff Rounce in Norfolk, a retired teacher, who had plenty of time to devote to his bees. His apiaries were the most organised I have seen anywhere, with all the hives on stands in woodland clearings and preserved with fresh creosote. He was equally meticulous with his management and it was obvious that his bees were given the best possible care. Jeff's wife was the village postmistress and undoubtedly some honey was sold from her Walsingham post office, but Jeff's major outlet was the Shrine Shop of Our Lady of Walsingham, the Shrine itself attracting over 10,000 visitors a year. Not surprisingly, Jeff depicted the Shrine's Anglican church on his honey labels, to good effect.

I could easily have sold honey in bulk to a small honey co-operative, but the price was very low. Tesco's was another possibility, for they were trying to promote local produce. They would pick up the honey from certain locations, take it to a central processing plant and then deliver it back to its supermarkets in the original area with the local county label pasted on the jars. Neither of these options were to my liking for I wanted complete control of the marketing of my honey. My aim – a very ambitious one – was to sell the honey to one of the top stores in London and throughout the whole of the UK, but how was this to be achieved? Val and I discussed this for many weeks and together we decided that we would set up our own honey business.

Way back in the 1950's Manley, one of Britain's finest beekeepers who had apiaries in the Chilterns, stated that in order for a beekeeper to be able to make a successful beekeeping business, three factors were of great importance: having the right amount of experience; having the capital needed; and having a good market for the hive produce. Essentially all three of these are needed concurrently. There was another factor to be considered by us, too; where we would find the time, for we both had responsible full-time jobs.

The only 'assets' that I could put into the honey business were my experience of keeping bees and selling honey on a small scale for about twenty years and what I thought to be some very good ideas. My plan was firstly to cut my colonies down to just fifty hives to give me more spare time and, secondly, to buy small amounts of many different types of honey, monofloral if possible, from elsewhere. I had already stayed with beekeepers in several different parts of Europe by then and very much liked the honeys which they produced.

I would then 'downpack' the imported honey – as well as my own – mainly into miniature 28g pots for which I would design my own packaging. The packages would then be sold either in 'up-market' food stores or by mail order as the latter, even before internet sales, was becoming increasingly popular for anniversaries and celebrations. By downpacking honey, although there would be more work, the honey would sell for a very high price and manageable amounts could be bought in as required. Hopefully, customers who tried our range of honeys would then want to buy full size jars of the honey varieties that they particularly liked.

Val had already decided on a business name, 'Honey Hunters', the very ambitious idea behind it being that eventually we would try and locate the source of any honey that was requested. We had an idea for the logo too: it would be based on a carved wooden bear we picked up in an antique shop in Goslar, Germany. As the bulk of the honey would be put into plastic 28g pots I decided to develop two types of packaging; one would hold twelve pots, the other thirty pots (a month's supply of honey for breakfast or tea time).

The smaller box was very simple. It would be rectangular with space for three rows of four pots; it would have a strong lid which could be removed and slipped under the box bottom so that the contents could be displayed; and there would be an inner, clear, perspex cover which would protect the pots.

The second type of packaging was much more complicated. I wanted it to be in the form of a dispenser which could be hung on the kitchen wall. It was to look attractive and resemble the pine tree from which the forest beekeepers cut out their combs from wild bees' nests. Each morning or afternoon, the recipient of the gift would be able to open the door in the tree hive and take out a pot of honey. As there were to be 30 types of honey in the dispenser, each day's honey would be of a different type and a surprise for the consumer – who would soon realise just how diverse honeys were in taste, aroma, colour and texture.

Once my ideas were on paper, complete with the appropriate measurements,

I contacted several box manufacturers before choosing one particular firm in Bradford for the work. They were extremely helpful throughout, except for one almost disastrous flaw at the end of the project. All the artwork – the logo design, the box decoration and tree motifs, as well as labels for the mini pots was done by a local bank manager's wife who was starting out in that line of business. We had already set up a list of honey types which we knew were available and provided the artist with a supply of pictures to help her with the work.

I needed cash for the project – about £10,000. I needed to go to the bank (Manley would no doubt have shuddered at this point). However, I had security, plenty of it: a secure job and a good property. And £10,000 didn't seem that much anyway; it was only the cost of a new car. One week after driving a new car you have lost a lot of money but I knew that my business plan looked good and I would make a healthy profit. The bank manager thought so too, even though he knew nothing at all about honey. He was either captivated by my enthusiasm or knew that the bank's money was going to be safe enough, given some security. The money would cover all the packaging and labels, the first batch of honey, the setting up of a honey house and equipping it, and a bulk supply of pots and jars in different sizes. It would also pay for the launch of the enterprise.

Before spending any money at all, I carefully weighed up the advantages/ disadvantages of being VAT registered. There was no doubt in my mind that although it would perhaps take years before I reached the threshold when I would be compelled to register, doing so straight away would be to my advantage. Not only would I receive the VAT back on all my purchases, this would extend to (though proportionally) my telephone bills, running my pick-up truck, etc, yet I would not have to charge and eventually repay VAT on my products as they were classed as food. One disadvantage, of course, was that there would be more paperwork and that returns would have to be made each quarter. However, I thought that doing the paperwork regularly would be a good discipline to keep on top of the business. I also employed an accountant who showed me how to keep my books, checked my VAT return and completed my annual tax return.

Just before I started on the design stage of the project, I had contacted several beekeepers who advertised in the magazine Abeilles de France and asked them to send me samples of their honey. The greatest and most interesting range of honeys came from Joel Schiro, a beekeeper in the Pyrenees, near Lourdes. As we had friends in Toulouse, we booked flights for a short holiday in the region and allowed ourselves three days with the beekeeper, who was just a couple of hours away by train. In that short time we learned a lot about French beekeeping, travelling from one large apiary to another, and spending time in the huge honey house, sampling even more varieties of honey. We put together a first order – just 15 x 35kg tubs and settled all the paperwork there and then. Each pail of honey contained clean, well-filtered honey just ready for bottling and some of the types we chose were honey from rhododendron (not the poisonous variety!), fir trees, bell heather (moorland), ling, sea holly, lime tree, sunflower, lavender, cherry, and spring mountain flowers. We also arranged transportation. A haulage firm across the road from us in our village made the trip down to Spain once every fortnight and would collect the two pallet loads on the way back – only adding an hour or so of driving for the slight diversion.

On our return to the UK we also ordered honey from British packers thus adding to our list of honeys: Mexican, Orange Blossom, Caribbean, New Zealand Clover, Tasmanian Leatherwood, Guatemalan, Chinese, Australian Bluebell and Canadian – all of which were available in 25kg tubs.

Within a short time we added many French 'herb' honeys to our list and manuka, from New Zealand, which was in great demand after Peter Nolan's work on this honey and its curative effect on stomach ulcers.

### My own 'honey house'

A purpose-built honey house was needed, but time and money were at a premium. Fortunately, I already had a concrete pad next to our house which would make a good base. It had formerly served as a base for a greenhouse made by the previous owner (I pulled it down as it was hideous – it was made of panels of perspex of different thicknesses and shapes gleaned from an old Lancaster aeroplane factory). I chose the easiest and cheapest solution, a pre-cast concrete garage 5m x 3m, but with a standard door fitted to the front instead of a full-width one for a car. I painted the inside walls and the floor, as the surfaces were very powdery, and made a false bee-proof roof. Both water and electricity were nearby so fitting these was not a problem.

Three banks of power points were fitted and all lights were shielded with plastic shades. I had a large galvanised sink, big enough for washing 35kg containers, made to my own design at a local metalworker's yard. One side of the honey house was for storage of large honey pails, an extractor and steel shelves for finished products, whilst opposite this was a bench for uncapping combs or bottling honey. Under the bench I had a row of honey warming boxes, bought from Thorne's, but extended in height with 'lifts' so that they would accommodate the larger tubs. I bought two pieces of equipment for the honey house: an electric zapper to kill flying insects and a Danish bottling machine which could be calibrated to fill even the very small 28gm pots.

The Honey House.

Before I used the honey house I asked the local environmental health officer to come and inspect the building. Legislation for honey processing rooms and their inspection was quite new at the time of the officer's visit. He knew nothing about honey except that it was in the 'low risk' category of foods. I soon realised that anyone fearing an inspection has no need to worry – as long as their honey-handling premises are clean and tidy. The important things were all in order; washing facilities were good (both hot and cold water), worktops, walls, ceilings and floors were all easy to clean, light bulbs were shielded, there was adequate light and ventilation (but who opens the honey house window during extracting?), electric points were safe and earthed, and the inspector liked the zapper and first aid kit which, apparently, weren't essential. I was given a score of 100%. I asked him how low the mark needed to be for an inspection to fail. He said if I had scored 60% he would have had to put me on his list for a visit in a year's time. Even hotels and restaurants in that category are given a chance to improve, but in a shorter space of time. Even then, though, the inspectors must make an appointment, they cannot make spot checks, so I was left thinking how limited the inspections were. I was amazed; this was the first time I had achieved a perfect score in anything!

One of the difficulties of introducing a new range of products onto the market is that you cannot hawk them around to shops until you finally have the finished products in your hands. It's no good going to retailers and saying 'I'm thinking of expanding my range of products. I'm going to put them in packs like this ... are you interested?' Nine times out of ten you will be told to come back later. I knew that there would be some time between having my products to show potential customers and sending them to food magazines and Sunday Supplements before I would get any return on my investment.

However, I had, I believed, one golden opportunity for launching the business – the Lincoln Christmas Market. Thousands of people from many parts of the UK, and abroad, are attracted to this annual event. From the picturesque castle and cathedral area and all down the Steep Hill crowds throng amongst the craft and food stalls. Food, beer and gifts from Germany, hot potatoes and chestnuts, French crepes, roast boar, speciality foods, Christmas Carols, Morris Men – the range of products and entertainment is enormous. What an opportunity! I paid my £200 rent space and constructed a market stall out of iron and plastic at the front of which would be an enormous trestle table that I would borrow from my school potting shed.

Ten days before the market I began to panic. The boxes I had ordered still hadn't been delivered. All the honey was waiting on shelves – hundreds of pots of 30 varieties of honey, all labelled and ready for packing. The filling of the small pots had taken a very long time for the Svienty machine either delivered plus or minus two or three grams each time; in order to work properly it needed a constant stream of honey at the same pressure. I had to resort to using a hand dispenser on a 28lb honey tin, which was a monotonous and painful job. After a series of phone calls I was told that the boxes would be brought the following evening. That was OK, I thought, as I still had one weekend before the market. The packaging looked superb when at last it arrived. When filled with a dozen varieties the rectangular presentation boxes looked really attractive. I then began to fill the 'Christmas Tree' dispenser- and realised very quickly that something was wrong – this box would only hold twenty-eight instead of the labelled '30 varieties'.

Our packaging.

I had a problem – or the company had. To this day, I believe they made the wrong measurements as they had the empty pots to take back to their design room. I suppose I could have, then and there, said – 'Take them back, they are not what I ordered'. But I had all the honey on the shelves and the prospect of a good market just a few days away. The director (he'd overseen the whole project from start to finish and had delivered the packages himself) absolved himself of any responsibility and said that he would have a new label produced for the following day and sent over by courier – and that this could be placed over the incorrect printing. I wasn't happy – I had planned for a 30-day supply of honey – and based my costings on 30 pots, but I was in a very difficult situation. I opted for the new label and I'm sure the director sighed with relief. A teaching colleague, who had himself been in business, knowing that I was under pressure, came one evening with his wife and helped us to fill the boxes ready for the market.

The Lincoln Christmas Market was a spectacular failure. I was unfortunate enough, like twenty other stall-holders, to have a very bad position in a small car park at the side of the castle. It was at this point that many people disembarked from their coaches and used the park as a thoroughfare. Many people stopped briefly for a cursory look and said that they would return later. Some commented on how pretty our products were and some said what a pity I wasn't selling jam! We stood and stamped our feet in the cold thick fog for one evening and two whole days hardly selling a thing. On the second day, the seller of baked potatoes sold all his potatoes to a man who had a better location in Castle Square; the latter had already sold what he thought would be enough potatoes for the three days. Both my wife and I joined the Breton pancake-maker from time to time, exchanging products instead of cash, strangely comforted by our joint disappointment in lack of trade.

And what had people bought chiefly at the Christmas market? The top seller was a Christmas hat carrying the title of a song, by Right Said Fred, which was then top of the pop charts, 'I'm too sexy'. Enough said!

Needless to say, we were totally numbed by our lack of success at what should have been an extremely lucrative market. If you can't sell gifts like the ones we had designed at Christmas time then the future of the whole business looked pretty bleak. In an attempt to get some cash back I managed to sell some boxes to some garden centres and up-market delicatessens, but local sales were very limited. I had already sent out gift packs for review in magazines and newspapers, yet I knew I couldn't afford to sit around and wait for the outcome; something more positive needed to be done and that quickly, for the hefty bank loan needed to be paid each month. For the short term, we placed some adverts in the Sunday newspapers saying that we would send our gift boxes directly by post to the intended recipients. The cost of each box, including package and postage was just £7.75 – cheaper by far than chocolates, flowers, or Stilton cheese! We had a good number of responses which was encouraging. However, we soon realised that this method of sale was only good if we persisted with our advertising – but the rates were really too expensive.

Our next line of attack was one that we had intended to do all along but had no time to do before Christmas, trying to get our gift boxes into one of the top London stores. We tried three. All of the buyers were impressed with our range and packaging – but in each case the answer was the same: 'We will get in touch with you'. Feeling flat, but not totally disappointed, we got on with more bottling and packing and I had time to produce a brochure which attempted to give interesting descriptions of each of the honeys we sold. The information included the geographic area the honey came from, their nectar sources, descriptions of the taste, colour, aroma and texture, and also some ideas how the various honeys could be used.

Within two or three months, several top food writers wrote favourable descriptions of our products and the market picked up a bit. Buying over the telephone was a problem for many customers as we were unable to accept credit card payment. In the case of emergency we sent out the product before a cheque arrived – and only once did we not receive payment.

One afternoon, unexpectedly and much to our delight, we received a telephone call from one of the prestigious London stores. They wanted to see us as they wished to order some of our products. Sadly, although they liked the gift boxes, they had something else in mind. They were building a new top-floor 'grocery' shop to compete with another famous nearby store. They wanted to stock all our range – but in one pound jars. Their requirements were specific as regards both the jars and labels. The jars had to be glass and square and the labels both plain and simple. The initial order was to be large and after that they expected to buy about 1,000 jars from us each month. We went home with a lot to think about. It was good to receive such an order, but it meant changing the way our business was intended to go. By down-packing honey into 28g pots I was getting a very good return on the 336g in each gift pack – some 150-200%. The problem was though, I wasn't selling enough of them. On the other hand, I would have the opportunity of selling my own two lines of honey in a top-notch store at twice the price that I could get locally. I thought, too, that it must be important to have a flexible approach to business; surely, being adaptable and taking opportunities when they crop up is a key to success?

We decided to go ahead with the store's requirements, except we had to compromise on the pots – we used square polycarbonate ones which were obtainable locally. Val was asked to go to London to train the staff on how to sell the honey and how to answer questions customers were most likely to ask such as 'Why does honey go solid?' and 'How long does honey keep for?' Also, not long after the store opened, we went to London again, this time to set up a honey-tasting stall in the shop. It was wonderful to see row after row of our square jars of honey with their distinctive yellow labels on the shelves of the grocery department, occupying a large and prominent position. We found, though, that our experience with customers confirmed our belief that the gift packs would sell. Whilst we found that the shoppers tasted some of the honeys and then went and bought a jar or two of the one pound sizes, most of them wanted to buy the gift packs which were not for sale. Many of the people who came to our stall were not the Saturday grocery shopping crowd, but tourists or business people looking for a gift to take home and our boxes were far more interesting and would travel better in suitcase than a jar of honey. We could not convince the floor manager at the end of the day – but at least we had managed to hand out some of our brochures which might attract some attention. On one of the tasting sessions, a gentleman with his whole family sampled honey after honey, eventually settling for one they all really liked. We pointed them in the direction of the shelves and they came back a few minutes later enthusiastically waving several pound jars. 'Oh, dear!' I said. 'You've picked up the wrong variety.' 'Yes! But these are the most expensive,' came the reply as they went off happily.

Eventually, the media really tuned in to our small business. Radio 4 wanted a sample box for their food programme and Carlton TV wanted to make a film about our honey. The BBC telephoned me just as I came in from work. They wanted the box the next day to be opened live on Derek Cooper's The Food Programme. I took a package straight away to the local village post office so that I could arrange for next day delivery. It was impossible. I was advised to go to the main post office in Gainsborough. I made it by 4.45pm. There was a long queue and out of the five positions behind the counter only two were manned. I don't know how I kept my patience – I knew that for guaranteed next day delivery the package had to be at the counter by 5pm. My turn came seven minutes after that – 'You're too late,' the woman said, 'It cannot go by Express post – you will just have to risk it getting there by First Class Post.' It was immutable. I was speechless! And cross, having waited for more than twenty minutes only to be disappointed. Fortunately, however, the honey arrived in time for Derek Cooper's programme, and received a favourable review, following which we had a fair number of enquiries from many parts of the UK.

The television programme was a success too. We had a very pleasant day with Tony Francis, who directed and produced Heart of the Country for Carlton TV, and his team. They came to our home in Lincolnshire and we looked into bee hives together, when unfortunately the novice reporter received a couple of stings despite excellent weather and a good nectar flow. Despite the pain, being a true professional, she continued with her commentary as soon as the stings had been removed. The team sampled the types of honey, on camera, as well as items of food and drink prepared by Val using monofloral types of honey. Interestingly, after the film was aired, we received requests for full-size jars of honey relating to the recipes Val used for the televised food – lavender honey for ice cream, dark fir tree honey for the fresh pineapple and yoghurt, lime tree honey for the honey lemonade, mild rhododendron honey for tea, and Caribbean honey for the rich, spicy walnut and honey cake. If people catch on to a recipe they will demand the right type of honey and in this way a market can be created.

The uniqueness of our honey business caught the attention of the staff of Farmer's Weekly, and Ann Rogers who wrote the farm life column came up to interview us together with a photographer. Subsequently, an article covering two full pages appeared in the weekly magazine, which not only covered our honey and beekeeping activities, but also my work locally as a rural science teacher. She also noted my criticism of MAFF policy which at the time restricted the movement of bees to control the spread of an invasive exotic pest. As the pest spread further from its original point of infestation in Devon, the Ministry gradually demarcated areas from which bees couldn't be moved. Since the mite, varroa, was going to spread eventually throughout the UK, as it had in almost every other country – and rapidly – despite restrictions on colony movements, the policy seemed senseless in a small country like Britain and meanwhile was going to make it very difficult for commercial bee farmers who needed to move bees for pollination contracts, or like myself and hundreds of other beekeepers, to take their bees to the heather moors. Ann also gave some of Val's hints for using honey in cookery, together with some of her recipes.

We had started our business at a very difficult time – in the middle of one of the worst recessions. However, with orders coming in from London and with slow but steady sales of gift boxes, we were quite hopeful about the future. We asked the London store how things were going – and they were pleased to say that we could confidently order more honey so that there would be no lack of continuity in supply. This was extremely important. Some of the honeys we procured from the Pyrenees and Provence could not be relied on every season and one of our main suppliers was going to cut down on some of the specialist mono-floral honeys as he had trouble selling them in France. He saw his future in putting most of his 3,000 hives of bees onto sunflowers and going for bulk sales. I had to buy more honey very quickly to keep in store – and in so doing, made my biggest, but unavoidable, mistake. I went back to the bank.

The bank saw no problem in supplying me with more cash to meet the honey bill. The manager had been a couple of times to our house, had seen what we were doing and was pleasantly surprised that we had one of England's most prestigious stores as a customer. I placed a large order with my French supplier. When I went to arrange collection with my haulage friend whose business was just across the road from me, I had a shock. Their lorries were no longer going to the southern border of France and Spain. As a result of minimum wage agreements, the catalogue supply company was no longer buying clothes and leather goods from Spain and Portugal as the labour costs were too high – they were now sourcing their goods in Asia. However, I was given a couple of contacts of hauliers that operated in that area and managed to get the honey, but at double the transportation cost. I continued to supply London with the honey but two months later they did not re-order – ever! We found out, although they did not tell us, that they were now procuring the honey themselves from the Pyrenees, bottling it in the cellar beneath the store, or one of their warehouses and selling it under their own label. Of course, we had no written contract with the company, for that is not their policy, despite the fact that, fool that I was, I had allowed them the distinction of being sole suppliers of our range in London. I learned that in the same week, a new floor manager had decided to have a change and got rid of a whole collection of bone china that someone had produced specifically for the shop, so I was not the only disappointed supplier.

Admittedly, one of my biggest mistakes had been to put most of my stock into one store. I should have heeded the advice of Val's uncle, a businessman, whose mantra was never to put more than 20-30% of your stock in one outlet. However, we were in a depression and times were hard and it had been too easy for me to swallow the bait. I knew of commercial beekeepers who were almost doubling their mileage each year in seeking out and supplying retail outlets as the local demand for their produce had diminished.

Out of the blue came one glimmer of hope. I received a fax from Malta for 1,000 jars of honey – and if they liked it they would order more. I had not contemplated exporting before and had a lot of paperwork to sort out before the deal could be completed. All was ready when I received another fax – apparently from the one of the food ministers in Malta – which requested a health certificate stating that the whole consignment of honey 'must not have come from any colonies which had had any disease of any type for the three years prior to harvesting'. This was an impossible request as all the problems, apart perhaps from foulbrood, are endemic in most apiaries throughout the world. I faxed my supplier. He had 3,000 colonies in many apiaries from the Atlantic seaboard to the Mediterranean. Three days later, much to my surprise I received a certificate of health from the French veterinary service endorsing the fact that all the colonies had been inspected and that they had been in excellent health over the last three years! The French authorities thought nothing of signing the certificate, no matter how dubious the declaration. My customer in Malta was pleased with the honey – and wanted more. However, unreasonable officialdom, or more likely pressure from a local honey supplier, made further exports impossible so I was still left with a huge amount of honey and no discernible market.

Eventually we decided that the only way that we could dispose of our honey was on a market stall. This was well before Farmer's Markets were in vogue, so we rented a space each week in the old indoor Butter Market in Newark, about 25 miles from our home. This meant we had to be preparing our items for sale on Friday evening and making an early start the next day to be at the market by 8am. For eighteen months, with Val's creative flair, we made the stall as attractive as possible and gave away free sample pots of new varieties to our regular customers and enjoyed talking to everyone who stopped by. Val made unique candles which she had covered with dried pressed flowers and tiny leaves that were then dipped once more in wax for a protective, transparent coat, which became very popular. Also, to increase our turnover, we combed market stalls, car boot sales and even antique shops for candlesticks – and old brass ones, complete with beeswax candles, were snapped up very quickly. A good range of cosmetics made by my friend Willie Robson's Chainbridge Honey Farm were an additional attraction – the peppermint foot cream, hand creams and lip balms all selling well.

Our market stall.

It was important, too, for us to use our time at Newark to educate people about bees. By trying our honeys, visitors to our stall soon realised that honey was not just sweet and sticky, but had a full range of tastes – fruity, spicy, and almost rank or bitter. Naturally, enquiries were made about what we added to our honey and when we answered, 'Nothing!' that the taste came from the flowers that the bees visited, there was incomprehension that the honey didn't taste for instance like the fruit of a plant. I took an observation hive to the stall each week so that our customers could watch the bees and learn about colony life and for many people this was the first opportunity they had of seeing bees close-up.

Whilst the days on the cold concrete market-floor were long, they seemed to pass by relatively quickly, for our regular customers always wanted to talk for a while and there were the other stall-holders with whom to chat. There, in the middle of the recession, they would work hard at selling their crafts, antiques, books and clothes though always, it seemed, the odds were against them. Each week they gave themselves a reason why they had made little money – it was the third week in the month and the people hadn't been paid yet; it was the beginning of the school term and money had to be spent on school uniform; it was summer and people were saving for their holidays: always there was an excuse. No matter what, they kept cheerful and confident, turning up each week and ever hopeful of a better time ahead. We didn't have their tenacity and when our range of products diminished and our debt was just about paid, we left the market forever. We also had a steady income from our full-time jobs, something most stallholders lacked; it was their main income.

It is easy to look back now and analyse what went wrong. Undoubtedly, we made mistakes but we were trying just too hard when market forces were against us and eventually we had no energy or wish to continue with our honey business. If internet shopping had been available at that time, we might have emerged from the depression with good postal sales, but after spending half a weekend for eighteen months selling honey, on top of our everyday jobs, we were tired and needed a rest. I still think our ideas were sound and if taken up by beekeepers today they would make a success of such a business, especially in tourist areas.

Have we regrets? Surprisingly few. We learned many aspects of developing and running a business and came into contact with people whom we would never have met otherwise, including someone who was inspired by our efforts to take up beekeeping and has made it a very successful livelihood. We are very open-minded people but not naive; we tend to trust people until they let us down. And we prefer it that way. We were able to utilise our creative and communication skills effectively, and of great importance, we made many, many friends. It was a short time of our lives that we look back on with nostalgia and not regret; and some of those contacts we made are still friends today.

From: A Beekeeper's Progress, by John Phipps.
Merlin Unwin, 2013

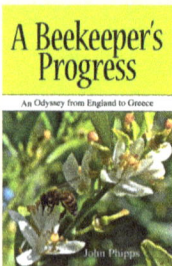

# Books Published by Northern Bee Books in 2019

**ON THE KEEPING OF BEES**
WHITAKER, JOHN M
ISBN: 9781912271481
WHITAKER 2ND EDITION. **£19.95**

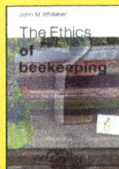

**THE ETHICS OF BEEKEEPING**
WHITAKER, JOHN M
ISBN: 9781912271245
HOOD 1ST EDITION. **£17.95**

**BEES AND MAN**
HOOD, WILLIAM MICHAEL
ISBN: 9781912271450

**MELINDA'S BEE HIVE: VOLUME 1**
DEMPSTER, SOULA
ISBN: 9781912271429
**£12.95**

**MELINDA'S BEE HIVE: VOLUME 2**
DEMPSTER, SOULA
ISBN: 9781912271436
**£12.95**

**NATUURIMKEREN MET DE WARRÉ-KAST. EEN HANDLEIDING**
HEAF, DAVID
ISBN: 9781912271412

**BIJENHOUDEN VOOR IEDEREEN**
WARRÉ, ABBÉ
ISBN: 9781912271344

**THE HONEYBEE IN FOCUS**
SWEENEY, WIN
ISBN: 9781912271375
**NEW £24.00**

**BEEKEEPING STUDY NOTES: BBKA CERTIFICATE IN BEEKEEPING HUSBANDARY**
YATES, J D
ISBN: 9781912271405
**£19.95**

# Books Published jointly by Northern Bee Books and IBRA in 2019

**THE HEALTHY HIVE GUIDE**
BASTERFIELD, DAN
ISBN: 9780860982890
**£16.95**

**AETHINA TUMIDA: UN PROBLEMA EMERGENTE NEL 21ESIMO SECOLO**
CARRECK, NORMAN L
ISBN: 9780860982791
**ITALIAN EDITION OF THE ENGLISH ED. £24.95**

**A BOOK OF HONEY**
CRANE, EVA
ISBN: 9780860982883
**REPRINT £29.95**

# Beekeeping Management Calendar

| MONTH | TASKS | FORAGE (Start . .) |
|---|---|---|
| SEPTEMBER | Treat colonies with two applications of Apiguard for varroa control. Prepare colonies for winter; feed strong syrup solution. Ensure colony has a fertile queen. | |
| OCTOBER | Continue feeding. Ensure hives are weather proof and in a sheltered position. Fit mouseguards and plastic skirts around the hives in woodpecker districts. Finish Winter BKQ. | Ivy |
| NOVEMBER DECEMBER JANUARY | Check apiaries frequently for storm damage, vandalism - or theft. Heft hives to assess the weight and therefore amount of stores. Feed blocks of candy if the colonies feel light. Check mouse guard is still in position or that it isn't blocked by dead bees. If there is a lot of snow, clear it from the roof - it is good insulation but if the sun is bright I slope boards in front of the hive to deter bees from flying as they will become chilled, fall in the snow and perish. Tidy up apiary. Melt down solid combs of oil seed rape honey. Make candles, wax polish, etc, for use in the home or to sell alongside honey at market stalls or craft fairs. Finish Spring BKQ. | Snowdrops |
| FEBRUARY | On mild days bees should be flying - check that pollen is being carried into the hive as this is a good sign that the queen is present and laying. At the end of the month, start feeding small amounts of syrup regularly to stimulate colony growth. | Heliotrope Winter Aconite Crocus Willows |
| MARCH | Provide bees with water - but not too close to the hive  - replenish until the end of summer. Check food supply and feed syrup if needed. More bees die of starvation at this time of year than any other. Get equipment ready for new season. Review last season's record cards and have them ready for use. Survey area for crops being grown. | Mahonia j. Daphne Fl. Currant Hazel Celandine |
| APRIL | First examinations can be made if temperature is over 15C -j ust a quick glimpse to see all is ok - winds are very cold in the East this month. Rape will be soon flowering in fields nearby; add supers well in advance of the colonies' needs. Inform farmers of colony locations and request timing of spraying operations. Finish summer BKQ. | Oil seed rape Dandelion Soft fruit Blackthorn |
| MAY | Swarming time from now until July. Start nine-day inspections for queen cells. Fit Snelgrove boards to colonies which show signs of swarming. Have equipment for collecting swarms ready - skep, sack, secateurs, small saw, Start extracting oil seed rape honey. Have bait hives in place (they should be completely dark, have a narrow entrance and be at least 3 - 4 metres above ground). | Top fruit Sycamore Deadnettles H. Chestnuts Holly Field beans |
| JUNE | Make nucleus colonies from queens reared above the Snelgrove Board and move to a new site. Give collected swarms and the nucleus colonies frames with new sheets of wax foundation and feed with syrup. Finish extracting of rape honey. Store, or leave on hives combs with solid honey. | Hawthorn Wild clover Dog rose |
| JULY | If colonies are to be moved to the heather, unite a nuc to a colony with an old queen. Put out wasp traps. | Lime Blackberry |
| August | Move colonies to the heather by the middle of the month. Ensure that they have plenty of food in the brood box. Fit supers with thin sheets of wax so that comb honey can be produced. All colonies apart from those going to the heather should have any surplus honey harvested by now. Unite colonies if no increase is desired or to fortify a colony with a new queen. Combs which are emptied of honey are sprayed with Certan to prevent wax moth damage and stored with a queen excluder at the top and bottom of the stacks to prevent mice or rats from getting at the combs. Finish Autumn BKQ. | Willowherb Water balsam Buddleia globosa |

# Floral Sources

| PLANT | COLOUR OF HONEY | YIELD | MAR | APR | MAY | JUN | JUL | AUG | SEP | OCT |
|---|---|---|---|---|---|---|---|---|---|---|
| IVY | GREENISH | E | | | | | | | | |
| MUSTARD | WHITE | A-C | | | | | | | | |
| LING | BROWN | D | | | | | | | | |
| BELL HEATHER | 'PORT WINE' | | | | | | | | | |
| PURPLE CLOVER | WATER WHITE | E | | | | | | | | |
| LIME | GREENISH WHITE 32-35% | D-F | | | | | | | | |
| WILLOWHERB | WATER WHITE | E-F | | | | | | | | |
| BLACKBERRY | LIGHT | | | | | | | | | |
| SWEET CHESTNUT | AMBER-RED | B | | | | | | | | |
| WHITE CLOVER | WHITE | C,D | | | | | | | | |
| BORAGE | WHITISH YELLOW 53% | D | | | | | | | | |
| RASPBERRY | WHITE 46% | C | | | | | | | | |
| FIELD BEANS | LIGHT TO VERY DARK 22% | C | | | | | | | | |
| HORSE CHESTNUT | LIGHT-DARK | D | | | | | | | | |
| SYCAMORE | GREENISH BROWN | D-F | | | | | | | | |
| HAWTHORN | LIGHT | B | | | | | | | | |
| WHITE DEADNETTLE | | D | | | | | | | | |
| OIL SEED RAPE | WATER WHITE | D | | | | | | | | |
| DANDELION | DEEP YELLOW | D | | | | | | | | |
| APPLE | LIGHT TO MEDIUM | B | | | | | | | | |
| CHERRY | LIGHT | B | | | | | | | | |
| PEAR | LIGHT | A | | | | | | | | |
| PLUM | LIGHT 15% | B | | | | | | | | |
| WILLOWS | LIGHT AMBER | D | | | | | | | | |

| POTENTIAL YIELD IN lb PER ACRE * | |
|---|---|
| A | 0-25 |
| B | 26-50 |
| C | 51-100 |
| D | 101-200 |
| E | 201-500 |
| F | over 500 |

*given the right weather and correct no. of colonies per acre.

Sugar concentration
Time of flowering

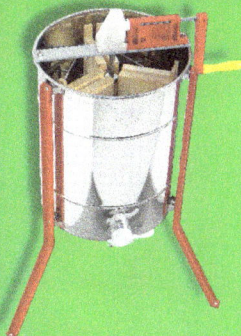

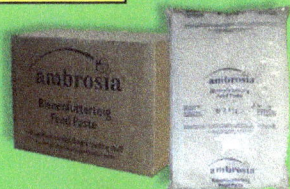

# Alphabetical Quiz

A       organism found in Malpighian tubules.

B       used to be a Gloucestershire beekeeping appliance dealer.

C       biological control for wax moth.

D       he explained honeybee parthenogenesis.

E       Chinese bee tree.

F       a good wasp plant.

G       honey bee tongue

H       gland supplying brood food.

I       Useful bee plant, but invasive

J       Corinthian peasant, Imperial and Royal Beekeeper.

K       beekeeping way, American, 9 frames in 10 frame hive.

L       bee plant, Custard and Cream.

M       UK city which featured bee sculptures in 2018

N       honeybee embryologist.

O       R. 'O.' B.M.

P       large skep promoter.

Q       biblically named American, responsible for Dadant frame.

R       US firm famous for its candles

S       bee disease, the larvae turn a greenish-yellow colour

T       honeybee mite.

U       couteau a desopercular;  Entdeckelungsmesser.

V       mid-gut of honeybee.

W       German, in 1883 he used artificial cell cups.

X       solvent for Canada balsam - ask a  microscopist.

Y       Metal spacing clip for side bars

Z       Nosema researcher.

Answers on Page 108

# 20

# DIARY & CALENDAR

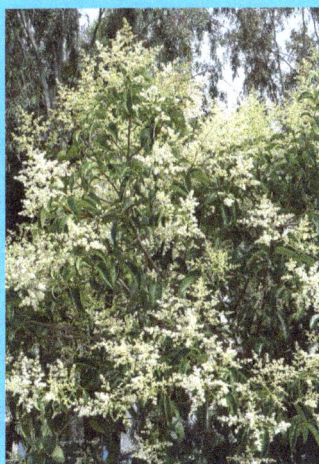

# - PART II -

## Some Interesting Greek Trees which May be Grown with Care in the UK.

**Edible loquat, *Eryobotrya japonica***
South walls, SW England and Ireland. Protection needed in severe winters.
Strongly scented but becomes rather ugly at fruiting time.
Loquats are a good source of calcium, potassium and fiber.
The bees work this tree in great numbers even at low temperatures.

| DAY | JANUARY 2019 FORAGE | TEMP | | WIND | | CL'D | RAIN | 1 | 2 | 3 |
|---|---|---|---|---|---|---|---|---|---|---|
| | | MIN | MAX | DIR | B.S | | | HIVE WEIGHT | | |
| 1 | | | | | | | | | | |
| 2 | | | | | | | | | | |
| 3 | | | | | | | | | | |
| 4 | | | | | | | | | | |
| 5 | | | | | | | | | | |
| 6 | | | | | | | | | | |
| 7 | | | | | | | | | | |
| 8 | | | | | | | | | | |
| 9 | | | | | | | | | | |
| 10 | | | | | | | | | | |
| 11 | | | | | | | | | | |
| 12 | | | | | | | | | | |
| 13 | | | | | | | | | | |
| 14 | | | | | | | | | | |
| 15 | | | | | | | | | | |
| 16 | | | | | | | | | | |
| 17 | | | | | | | | | | |
| 18 | | | | | | | | | | |
| 19 | | | | | | | | | | |
| 20 | | | | | | | | | | |
| 21 | | | | | | | | | | |
| 22 | | | | | | | | | | |
| 23 | | | | | | | | | | |
| 24 | | | | | | | | | | |
| 25 | | | | | | | | | | |
| 26 | | | | | | | | | | |
| 27 | | | | | | | | | | |
| 28 | | | | | | | | | | |
| 29 | | | | | | | | | | |
| 30 | | | | | | | | | | |
| 31 | | | | | | | | | | |

# JAN20

| | |
|---|---|
| | **8,WE** |
| **1,WE**<br>NEW YEAR'S DAY<br>ST BASIL (ORTHODOX) | **9,TH**<br>LIBERATION DAY (GUERNSEY, JERSEY) |
| **2,TH**<br>BANK HOLIDAY SCOTLAND | **10,FR** |
| **3,FR** | 11,SA |
| 4,SA | 12,SU |
| 5,SU | **13,MO** |
| **6,MO**<br>EPIPHANY | **14,TU** |
| **7,TU**<br>ORTHODOX CHRISTMAS | **15,WE** |

| | |
|---|---|
| **16,TH** | **24,FR** |
| **17,FR** | 25,SA<br>BURN'S NIGHT |
| 18,SA | 26,SU |
| 19,SU | **27,MO** |
| **20,MO** | **28,TU** |
| **21,TU** | **29,WE** |
| **22,WE** | **30,TH** |
| **23,TH** | **31,FR** |

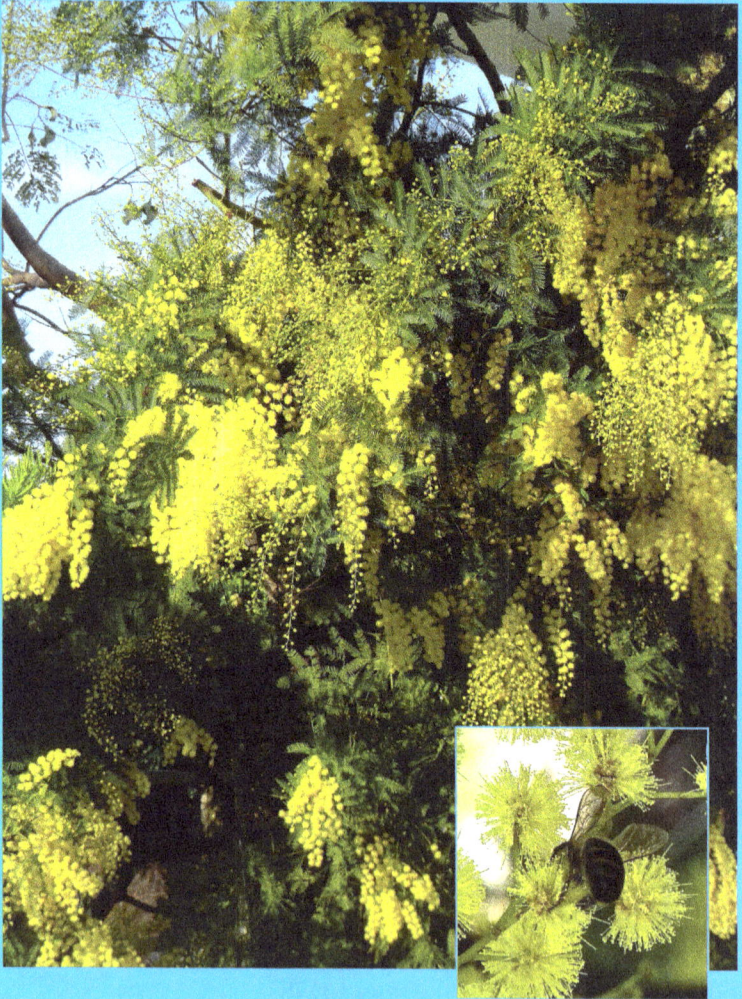

**Mimosa, *Acacia dealbata***
*Often grown in greenhouses, outdoors it is wind tolerant and can survive temperatures as low as -5C. Horticultural fleece on the branches helps in frosty weather as well as a good mulch of dried leaves to protect the roots. Heavily scented.*

| DAY | FEBRUARY 2019 FORAGE | TEMP MIN | MAX | WIND DIR | B.S | CL'D | RAIN | 1 | 2 | 3 HIVE WEIGHT |
|---|---|---|---|---|---|---|---|---|---|---|
| 1 | | | | | | | | | | |
| 2 | | | | | | | | | | |
| 3 | | | | | | | | | | |
| 4 | | | | | | | | | | |
| 5 | | | | | | | | | | |
| 6 | | | | | | | | | | |
| 7 | | | | | | | | | | |
| 8 | | | | | | | | | | |
| 9 | | | | | | | | | | |
| 10 | | | | | | | | | | |
| 11 | | | | | | | | | | |
| 12 | | | | | | | | | | |
| 13 | | | | | | | | | | |
| 14 | | | | | | | | | | |
| 15 | | | | | | | | | | |
| 16 | | | | | | | | | | |
| 17 | | | | | | | | | | |
| 18 | | | | | | | | | | |
| 19 | | | | | | | | | | |
| 20 | | | | | | | | | | |
| 21 | | | | | | | | | | |
| 22 | | | | | | | | | | |
| 23 | | | | | | | | | | |
| 24 | | | | | | | | | | |
| 25 | | | | | | | | | | |
| 26 | | | | | | | | | | |
| 27 | | | | | | | | | | |
| 28 | | | | | | | | | | |

# FEB20

| | |
|---|---|
| **FEB20** | 8,SA |
| 1,SA | 9,SU |
| 2,SU<br>CANDLEMAS | 10,MO |
| 3,MO ○ | 11,TU<br>ST GOBNAIT,<br>PATRON SAINT OF IRISH BEEKEEPERS |
| 4,TU | 12,WE |
| 5,WE<br>CHINESE NEW YEAR | 13,TH<br>ST MODMONOC, IRELAND |
| 6,TH | 14,FR<br>ST VALENTINE |
| 7,FR | 15,SA |

| | |
|---|---|
| 16,SU | 24,MO |
| 17,MO | 25,TU<br>SHROVE TUESDAY |
| 18,TU | 26,WE<br>ASH WEDNESDAY |
| 19,WE | 27,TH |
| 20,TH | 28,FR |
| 21,FR | 29,SA |
| 22,SA | |
| 23,SU<br>ST KHALAMPII, PATRON SAINT OF BULGARIAN<br>BEEKEEPERS (HIVE-SHAPED PIES BAKED) | |

**Judas Tree, *Cercis siliquastrum***
*Grows in a wide range of soils, in almost any position.*
*Very hardy, excellent specimen tree with its vivid flowers and strong scent.*
*The flowers grow directly from the branches of the tree.*

| DAY | MARCH 2019 / FORAGE | TEMP MIN | MAX | WIND DIR | B.S | CL'D | RAIN | 1 | 2 | 3 HIVE WEIGHT |
|---|---|---|---|---|---|---|---|---|---|---|
| 1 | | | | | | | | | | |
| 2 | | | | | | | | | | |
| 3 | | | | | | | | | | |
| 4 | | | | | | | | | | |
| 5 | | | | | | | | | | |
| 6 | | | | | | | | | | |
| 7 | | | | | | | | | | |
| 8 | | | | | | | | | | |
| 9 | | | | | | | | | | |
| 10 | | | | | | | | | | |
| 11 | | | | | | | | | | |
| 12 | | | | | | | | | | |
| 13 | | | | | | | | | | |
| 14 | | | | | | | | | | |
| 15 | | | | | | | | | | |
| 16 | | | | | | | | | | |
| 17 | | | | | | | | | | |
| 18 | | | | | | | | | | |
| 19 | | | | | | | | | | |
| 20 | | | | | | | | | | |
| 21 | | | | | | | | | | |
| 22 | | | | | | | | | | |
| 23 | | | | | | | | | | |
| 24 | | | | | | | | | | |
| 25 | | | | | | | | | | |
| 26 | | | | | | | | | | |
| 27 | | | | | | | | | | |
| 28 | | | | | | | | | | |
| 29 | | | | | | | | | | |
| 30 | | | | | | | | | | |
| 31 | | | | | | | | | | |

# MAR20

| | |
|---|---|
| | **8,SU**<br>NTERNATIONAL WOMEN'S DAY |
| **1,SU**<br>ST. DAVID'S DAY (WALES) | **9,MO** |
| **2,MO**<br>CLEAN MONDAY (ORTHODOX) | **10,TU** |
| **3,TU** | **11,WE** |
| **4,WE** | **12,TH** |
| **5,TH** | **13,FR** |
| **6,FR** | 14,SA |
| 7,SA | 15,SU |

| | |
|---|---|
| **16,MO** | **24,TU** |
| **17,TU**<br>ST PATRICK'S DAY | **25,WE** |
| **18,WE** | **26,TH** |
| **19,TH** | **27,FR** |
| **20,FR**<br>SPRING EQUINOX | 28,SA |
| 21,SA | 29,SU<br>BST BEGINS |
| 22,SU<br>MOTHER'S DAY | **30,MO**<br>ST ALEXIUS DAY (UKRAINIAN BEEKEEPERS HANG ICONS OF THEIR PATRON SAINTS OF BEEKEEPING, ST SAVVATY AND ST ZOSIMA IN SHRINES AMONGST THEIR HIVES) |
| **23,MO**<br>GREEK INDEPENDENCE DAY | **31, TU** |

**False Acacia, *Robinia psuedoacacia***
*Vigorous hardy tree which spreads easily by suckers and seeds,*
*considered by some as a nuisance.*
*Sheltered position is best, but tolerant of most soils.*
*Wonderful perfume.*

| DAY | APRIL 2019 FORAGE | TEMP MIN | MAX | WIND DIR | B.S | CL'D | RAIN | 1 | 2 | 3 |
|-----|-------------------|----------|-----|----------|-----|------|------|---|---|---|
| | | | | | | | | HIVE WEIGHT | | |
| 1 | | | | | | | | | | |
| 2 | | | | | | | | | | |
| 3 | | | | | | | | | | |
| 4 | | | | | | | | | | |
| 5 | | | | | | | | | | |
| 6 | | | | | | | | | | |
| 7 | | | | | | | | | | |
| 8 | | | | | | | | | | |
| 9 | | | | | | | | | | |
| 10 | | | | | | | | | | |
| 11 | | | | | | | | | | |
| 12 | | | | | | | | | | |
| 13 | | | | | | | | | | |
| 14 | | | | | | | | | | |
| 15 | | | | | | | | | | |
| 16 | | | | | | | | | | |
| 17 | | | | | | | | | | |
| 18 | | | | | | | | | | |
| 19 | | | | | | | | | | |
| 20 | | | | | | | | | | |
| 21 | | | | | | | | | | |
| 22 | | | | | | | | | | |
| 23 | | | | | | | | | | |
| 24 | | | | | | | | | | |
| 25 | | | | | | | | | | |
| 26 | | | | | | | | | | |
| 27 | | | | | | | | | | |
| 28 | | | | | | | | | | |
| 29 | | | | | | | | | | |
| 30 | | | | | | | | | | |

## APR20

**8,WE**

**1,WE**

**9,TH**
MAUNDY THURSDAY

**2,TH**

**10,FR**
GOOD FRIDAY

**3,FR**

11,SA

4,SA

12,SU
EASTER DAY
PALM SUNDAY (ORTHODOX)

5,SU
PALM SUNDAY

**13,MO**
EASTER MONDAY

**6,MO**

**14,TU**

**7,TU**

**15,WE**

| | |
|---|---|
| **16,TH** | **24,FR**<br>RAMADAN BEGINS |
| **17,FR**<br>MEGALI PARASKEVI (ORTHODOX) | 25,SA<br>ANZAC DAY |
| 18,SA | 26,SU |
| 19,SU<br>PASCHA (ORTHODOX EASTER) | **27,MO** |
| **20,MO**<br>EASTER MONDAY (ORTHODOX) | **28,TU** |
| **21,TU** | **29,WE** |
| **22,WE** | **30,TH**<br>ST ZOSIMA - 'GREET THE BEE ON ZOSIMA'S DAY AND THERE WILL BE HIVES AND WAX'. |
| **23,TH**<br>ST. GEORGE'S DAY<br>SHAKESPEARE DAY | |

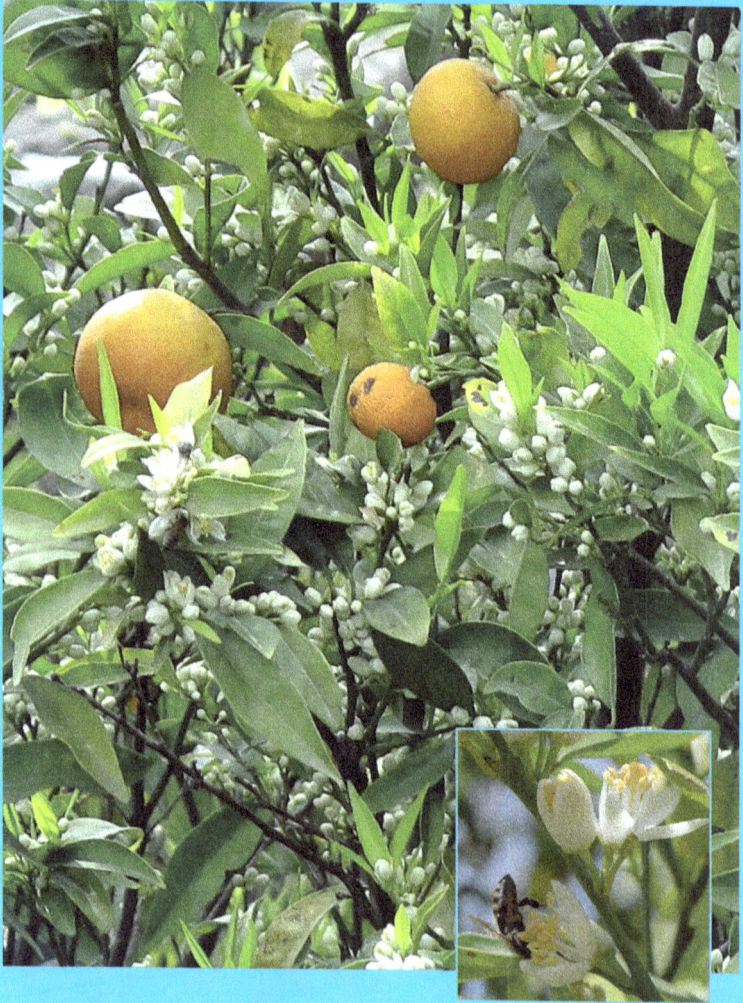

**Citrus Trees**

Many citrus trees can be grown in large pots and brought out once the temperature is above 10C. Hopefully, the pots will be outside at the time of flowering so that the scent of the flowers and the activities of the bees add enchantment to a garden. Kumquats can also be grown and they are hardier then other citrus species.

| DAY | MAY 2019 FORAGE | TEMP MIN | TEMP MAX | WIND DIR | WIND B.S | CL'D | RAIN | 1 | 2 | 3 HIVE WEIGHT |
|-----|-----------------|----------|----------|----------|----------|------|------|---|---|---------------|
| 1 | | | | | | | | | | |
| 2 | | | | | | | | | | |
| 3 | | | | | | | | | | |
| 4 | | | | | | | | | | |
| 5 | | | | | | | | | | |
| 6 | | | | | | | | | | |
| 7 | | | | | | | | | | |
| 8 | | | | | | | | | | |
| 9 | | | | | | | | | | |
| 10 | | | | | | | | | | |
| 11 | | | | | | | | | | |
| 12 | | | | | | | | | | |
| 13 | | | | | | | | | | |
| 14 | | | | | | | | | | |
| 15 | | | | | | | | | | |
| 16 | | | | | | | | | | |
| 17 | | | | | | | | | | |
| 18 | | | | | | | | | | |
| 19 | | | | | | | | | | |
| 20 | | | | | | | | | | |
| 21 | | | | | | | | | | |
| 22 | | | | | | | | | | |
| 23 | | | | | | | | | | |
| 24 | | | | | | | | | | |
| 25 | | | | | | | | | | |
| 26 | | | | | | | | | | |
| 27 | | | | | | | | | | |
| 28 | | | | | | | | | | |
| 29 | | | | | | | | | | |
| 30 | | | | | | | | | | |
| 31 | | | | | | | | | | |

# MAY20

| | |
|---|---|
| **1,FR** | **8,FR** |
| **2,SA**<br>MAY BANK HOLIDAY | **9,SA**<br>LIBERATION DAY (GUERNSEY, JERSEY) |
| 3,SU | 10,SU |
| **4,MO** | **11,MO** |
| **5,TU** | **12,TU** |
| **6,WE** | **13,WE** |
| **7,TH** | **14,TH** |
| | **15,FR** |

| | |
|---|---|
| 16,SA | 24,SU |
| 17,SU | 25,MO<br>SPRING BANK HOLIDAY |
| 18,MO | 26,TU |
| 19,TU | 27,WE |
| 20,WE<br>WORLD BEE DAY | 28,TH |
| 21,TH<br>ASCENSION DAY | 29,FR |
| 22,FR | 30,SA |
| 23,SA | 31,SU<br>PENTECOST |

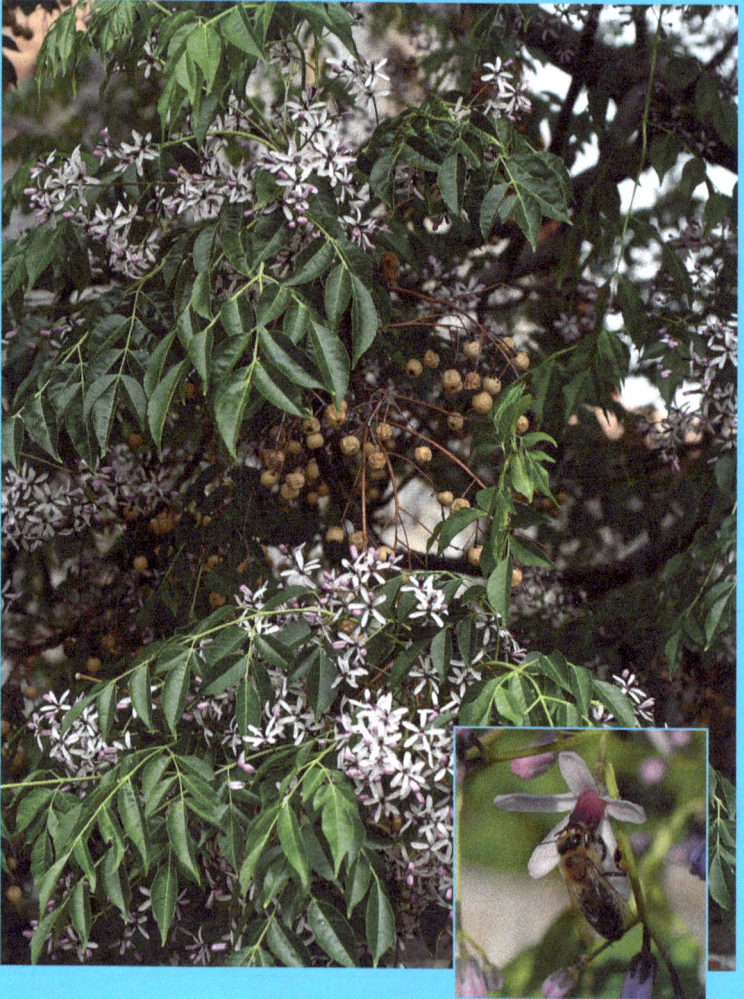

**Persian Lilac,** *Melia azedarach*
Often called the bead tree since for most of the year the pale orange berries
cover the branches. Its bluey-lilac flowers have a wonderful scent but,
unfortunately, the berries scatter everywhere and grow wherever they settle.

| DAY | JUNE 2019 FORAGE | TEMP | | WIND | | CL'D | RAIN | 1 | 2 | 3 |
|---|---|---|---|---|---|---|---|---|---|---|
| | | MIN | MAX | DIR | B.S | | | HIVE WEIGHT | | |
| 1 | | | | | | | | | | |
| 2 | | | | | | | | | | |
| 3 | | | | | | | | | | |
| 4 | | | | | | | | | | |
| 5 | | | | | | | | | | |
| 6 | | | | | | | | | | |
| 7 | | | | | | | | | | |
| 8 | | | | | | | | | | |
| 9 | | | | | | | | | | |
| 10 | | | | | | | | | | |
| 11 | | | | | | | | | | |
| 12 | | | | | | | | | | |
| 13 | | | | | | | | | | |
| 14 | | | | | | | | | | |
| 15 | | | | | | | | | | |
| 16 | | | | | | | | | | |
| 17 | | | | | | | | | | |
| 18 | | | | | | | | | | |
| 19 | | | | | | | | | | |
| 20 | | | | | | | | | | |
| 21 | | | | | | | | | | |
| 22 | | | | | | | | | | |
| 23 | | | | | | | | | | |
| 24 | | | | | | | | | | |
| 25 | | | | | | | | | | |
| 26 | | | | | | | | | | |
| 27 | | | | | | | | | | |
| 28 | | | | | | | | | | |
| 29 | | | | | | | | | | |
| 30 | | | | | | | | | | |

| | |
|---|---|
| **JUN20** | **8,MO** |
| **1,MO** | **9,TU** |
| **2,TU** ○ | **10,WE** |
| **3,WE** | **11,TH**<br>CORPUS CHRISTI |
| **4,TH** | **12,FR** |
| **5,FR** | 13,SA |
| 6,SA | 14,SU |
| 7,SU<br>PENTECOST (ORTHODOX)<br>TRINITY SUNDAY | **15,MO** |

| | |
|---|---|
| **16,TU** | **24,WE** |
| **17,WE** | **25,TH** |
| **18,TH** | **26,FR** |
| **19,FR** | 27,SA |
| 20,SA | 28,SU |
| 21,SU<br>SUMMER SOLSTICE | **29,MO** |
| **22,MO**<br>WINDRUSH DAY | **30,TU** |
| **23,TU** | |

# JULY

### Jacaranda mimosifolia

*This spectacular tree with its blue flowers and strong perfume stands out really well for specimen planting, especially if in an open position and seen against a pale blue sky. It is a long shot for planting in most gardens in the UK as some authors remark that it will only tolerate a tiny amount of frost. Others, however, say that it can survive in temperatures down to -5C, but this will affect the number of flowers which are produced. Kept trimmed, it may be possible to use fleece to give the tree protection. With very hot summers becoming the norm, jacarandas thrive in very high temperatures.*

| DAY | JULY 2019 FORAGE | TEMP MIN | MAX | WIND DIR | B.S | CL'D | RAIN | 1 | 2 | 3 |
|-----|------------------|----------|-----|----------|-----|------|------|---|---|---|
| | | | | | | | | HIVE WEIGHT | | |
| 1 | | | | | | | | | | |
| 2 | | | | | | | | | | |
| 3 | | | | | | | | | | |
| 4 | | | | | | | | | | |
| 5 | | | | | | | | | | |
| 6 | | | | | | | | | | |
| 7 | | | | | | | | | | |
| 8 | | | | | | | | | | |
| 9 | | | | | | | | | | |
| 10 | | | | | | | | | | |
| 11 | | | | | | | | | | |
| 12 | | | | | | | | | | |
| 13 | | | | | | | | | | |
| 14 | | | | | | | | | | |
| 15 | | | | | | | | | | |
| 16 | | | | | | | | | | |
| 17 | | | | | | | | | | |
| 18 | | | | | | | | | | |
| 19 | | | | | | | | | | |
| 20 | | | | | | | | | | |
| 21 | | | | | | | | | | |
| 22 | | | | | | | | | | |
| 23 | | | | | | | | | | |
| 24 | | | | | | | | | | |
| 25 | | | | | | | | | | |
| 26 | | | | | | | | | | |
| 27 | | | | | | | | | | |
| 28 | | | | | | | | | | |
| 29 | | | | | | | | | | |
| 30 | | | | | | | | | | |
| 31 | | | | | | | | | | |

# JUL20

| | |
|---|---|
| | **8,WE** |
| **1,WE** | **9,TH** |
| **2,TH** | **10,FR** |
| **3,FR** | 11,SA |
| 4,SA | 12,SU<br>BATTLE OF THE BOYNE (NORTHERN IRELAND) |
| 5,SU | **13,MO** |
| **6,MO** | **14,TU** |
| **7,TU** | **15,WE** |

| | |
|---|---|
| **16,TH** | **24,FR** |
| **17,FR** | 25,SA |
| 18,SA | 26,SU |
| 19,SU | **27,MO** |
| **20,MO** | **28,TU** |
| **21,TU** | **29,WE** |
| **22,WE** | **30,TH** |
| **23,TH** | **31,FR** |

**Eucalyptus**

For UK planting, the RHS recommends *E. pauciflora*, *E. gunnii*, *E. dalrympleana* and *E. parvula*. It is best to check which species are best suited to your location. Whilst the flowers have a wonderful perfume, it is the turps-like smell which comes from the trees in very hot weather which makes one aware of why the oil has been used in medicine. As regards flowering, the trees are fickle. Whilst flower buds are abundant on the trees it can be a long while before they break into flower. Unfortunately, in some parts of Europe the trees are devastated by lurp psyllids which are recognized by the little white houses (lurps) they secrete over themselves for protection. There is much antipathy about the planting of many eucalypts close together as they use huge amounts of water and they are recognised as a fire hazard.

| DAY | AUGUST 2019 FORAGE | TEMP MIN | TEMP MAX | WIND DIR | WIND B.S | CL'D | RAIN | 1 | 2 | 3 HIVE WEIGHT |
|---|---|---|---|---|---|---|---|---|---|---|
| 1 | | | | | | | | | | |
| 2 | | | | | | | | | | |
| 3 | | | | | | | | | | |
| 4 | | | | | | | | | | |
| 5 | | | | | | | | | | |
| 6 | | | | | | | | | | |
| 7 | | | | | | | | | | |
| 8 | | | | | | | | | | |
| 9 | | | | | | | | | | |
| 10 | | | | | | | | | | |
| 11 | | | | | | | | | | |
| 12 | | | | | | | | | | |
| 13 | | | | | | | | | | |
| 14 | | | | | | | | | | |
| 15 | | | | | | | | | | |
| 16 | | | | | | | | | | |
| 17 | | | | | | | | | | |
| 18 | | | | | | | | | | |
| 19 | | | | | | | | | | |
| 20 | | | | | | | | | | |
| 21 | | | | | | | | | | |
| 22 | | | | | | | | | | |
| 23 | | | | | | | | | | |
| 24 | | | | | | | | | | |
| 25 | | | | | | | | | | |
| 26 | | | | | | | | | | |
| 27 | | | | | | | | | | |
| 28 | | | | | | | | | | |
| 29 | | | | | | | | | | |
| 30 | | | | | | | | | | |
| 31 | | | | | | | | | | |

# AUG20

| | |
|---|---|
| | **8,SA** |
| **1,SA** | **9,SU** |
| **2,SU** | **10,MO** |
| **3,MO** ○ | **11,TU** |
| **4,TU** | **12,WE** |
| **5,WE**<br>SUMMER BANK HOLIDAY (SCOTLAND) | **13,TH** |
| **6,TH** | **14,FR**<br>THE SAVIOR OF THE HONEY FEAST (UKRAINE) |
| **7,FR** | **15,SA**<br>ASSUMPTION OF THE VIRGIN<br>'APOKIMISIS TIS PANAGIAS' |

| | |
|---|---|
| 16,SU | **24,MO**<br>**ST BARTHOLOMEW (TRADITIONAL DAY FOR**<br>**HARVESTING HONEY)** |
| **17,MO** | **25,TU** |
| **18,TU** | **26,WE** |
| **19,WE** | **27,TH** |
| **20,TH**<br>NATIONAL HONEY BEE DAY (USA) | **28,FR** |
| **21,FR** | 29,SA |
| 22,SA | 30,SU |
| 23,SU | **31,MO**<br>**SUMMER BANK HOLIDAY (ENGLAND, WALES,**<br>**NORTHERN IRELAND, GUERNSEY & JERSEY)** |

**Bottlebrush, *Callistemon spp.***
These brightly-coloured flowers add a lot of interest to a garden with their very long slender stamens which crowd together, so much so that bees look as if they are trapped within them as they try and reach the nectaries. The plants can be grown in the milder parts of the UK, notably where temperatures do not fall lower than -5C. Although they prefer an open sunny position, they benefit from being grown in a sheltered spot facing south or west.

| DAY | SEPTEMBER 2019 FORAGE | TEMP MIN | TEMP MAX | WIND DIR | WIND B.S | CL'D | RAIN | 1 | 2 | 3 HIVE WEIGHT |
|---|---|---|---|---|---|---|---|---|---|---|
| 1 | | | | | | | | | | |
| 2 | | | | | | | | | | |
| 3 | | | | | | | | | | |
| 4 | | | | | | | | | | |
| 5 | | | | | | | | | | |
| 6 | | | | | | | | | | |
| 7 | | | | | | | | | | |
| 8 | | | | | | | | | | |
| 9 | | | | | | | | | | |
| 10 | | | | | | | | | | |
| 11 | | | | | | | | | | |
| 12 | | | | | | | | | | |
| 13 | | | | | | | | | | |
| 14 | | | | | | | | | | |
| 15 | | | | | | | | | | |
| 16 | | | | | | | | | | |
| 17 | | | | | | | | | | |
| 18 | | | | | | | | | | |
| 19 | | | | | | | | | | |
| 20 | | | | | | | | | | |
| 21 | | | | | | | | | | |
| 22 | | | | | | | | | | |
| 23 | | | | | | | | | | |
| 24 | | | | | | | | | | |
| 25 | | | | | | | | | | |
| 26 | | | | | | | | | | |
| 27 | | | | | | | | | | |
| 28 | | | | | | | | | | |
| 29 | | | | | | | | | | |
| 30 | | | | | | | | | | |

| SEP20 | 8,TU |
|---|---|
| 1,TU | 9,WE |
| 2,WE | 10,TH |
| 3,TH | 11,FR |
| 4,FR | 12,SA |
| 5,SA | 13,SU |
| 6,SU | 14,MO |
| 7,MO | 15,TU |

| | |
|---|---|
| **16,WE** | **24,TH** |
| **17,TH** | **25,FR** |
| **18,FR** | 26,SA |
| 19,SA | 27,SU |
| 20,SU | **28,MO**<br>YOM KIPPUR |
| **21,MO** | **29, TU** |
| **22,TU** | **30, WE** |
| **23,WE**<br>AUTUMN EQUINOX | |

**Carob,** *Ceratonia siliqua*

This plant produces female and male flowers on different trees. The very strong smelling catkin-like flowers produce masses of pollen and, like the Judas tree, the flowers grow directly from the branches. It is undoubtedly one of the best pollen producing plants which allows colonies in Greece to build up after the summer drought and decreasing hive populations. The long beans produced on the female trees are used in Greece for feeding to livestock, though elsewhere in the world they are utilised as a chocolate substitute.

The trees are hardy in mild parts of the UK though prefer a sunny position facing south or west.

| DAY | OCTOBER 2019 FORAGE | TEMP MIN | MAX | WIND DIR | B.S | CL'D | RAIN | 1 | 2 | 3 HIVE WEIGHT |
|---|---|---|---|---|---|---|---|---|---|---|
| 1 | | | | | | | | | | |
| 2 | | | | | | | | | | |
| 3 | | | | | | | | | | |
| 4 | | | | | | | | | | |
| 5 | | | | | | | | | | |
| 6 | | | | | | | | | | |
| 7 | | | | | | | | | | |
| 8 | | | | | | | | | | |
| 9 | | | | | | | | | | |
| 10 | | | | | | | | | | |
| 11 | | | | | | | | | | |
| 12 | | | | | | | | | | |
| 13 | | | | | | | | | | |
| 14 | | | | | | | | | | |
| 15 | | | | | | | | | | |
| 16 | | | | | | | | | | |
| 17 | | | | | | | | | | |
| 18 | | | | | | | | | | |
| 19 | | | | | | | | | | |
| 20 | | | | | | | | | | |
| 21 | | | | | | | | | | |
| 22 | | | | | | | | | | |
| 23 | | | | | | | | | | |
| 24 | | | | | | | | | | |
| 25 | | | | | | | | | | |
| 26 | | | | | | | | | | |
| 27 | | | | | | | | | | |
| 28 | | | | | | | | | | |
| 29 | | | | | | | | | | |
| 30 | | | | | | | | | | |
| 31 | | | | | | | | | | |

| OCT20 | 8,TH |
|---|---|
| 1,TH ○ | 9,FR |
| 2,FR | 10,SA |
| 3,SA | 11,SU |
| 4,SU | 12,MO |
| 5,MO | 13,TU |
| 6,TU | 14,WE |
| 7,WE | 15,TH |

| | |
|---|---|
| **16,FR** | 24,SA |
| 17,SA | 25,SU<br>BST ENDS |
| 18,SU | **26,MO** |
| **19,MO** | **27,TU** |
| **20,TU** | **28,WE**<br>'OHI DAY' GREECE ('NO' TO MUSSOLINI)<br>THANKSGIVING DAY, USA |
| **21,WE** | **29,TH** |
| **22,TH** | **30,FR** |
| **23,FR** | 31,SA<br>HALLOWE'EN |

**Heather, *Erica verticallata***
This species of heather flowers in the first months of winter in the mountains of Greece.
It would be a late source of forage for bees in the UK, however, there seems to be
some confusion about its name. Any search shows information only for *Erica verticillata*,
a South African species, which has become extinct in the wild.

| DAY | NOVEMBER 2019 FORAGE | TEMP MIN | MAX | WIND DIR | B.S | CL'D | RAIN | 1 | 2 | 3 |
|-----|--------------------|------|-----|------|-----|------|------|---|---|---|
| | | | | | | | | HIVE WEIGHT | | |
| 1 | | | | | | | | | | |
| 2 | | | | | | | | | | |
| 3 | | | | | | | | | | |
| 4 | | | | | | | | | | |
| 5 | | | | | | | | | | |
| 6 | | | | | | | | | | |
| 7 | | | | | | | | | | |
| 8 | | | | | | | | | | |
| 9 | | | | | | | | | | |
| 10 | | | | | | | | | | |
| 11 | | | | | | | | | | |
| 12 | | | | | | | | | | |
| 13 | | | | | | | | | | |
| 14 | | | | | | | | | | |
| 15 | | | | | | | | | | |
| 16 | | | | | | | | | | |
| 17 | | | | | | | | | | |
| 18 | | | | | | | | | | |
| 19 | | | | | | | | | | |
| 20 | | | | | | | | | | |
| 21 | | | | | | | | | | |
| 22 | | | | | | | | | | |
| 23 | | | | | | | | | | |
| 24 | | | | | | | | | | |
| 25 | | | | | | | | | | |
| 26 | | | | | | | | | | |
| 27 | | | | | | | | | | |
| 28 | | | | | | | | | | |
| 29 | | | | | | | | | | |
| 30 | | | | | | | | | | |

# NOV20

| | |
|---|---|
| | **8,SU** |
| **1,SU**<br>ALL SAINTS DAY | **9,MO** |
| **2,MO**<br>ALL SOULS DAY | **10,TU**<br>REMEMBRANCE SUNDAY |
| **3,TU** | **11,WE** |
| **4,WE** | **12,TH** |
| **5,TH**<br>GUY FAWKES DAY | **13,FR** |
| **6,FR** | **14,SA** |
| **7,SA** | **15,SU** |

| | |
|---|---|
| **16,MO** | **24,TU** |
| **17,TU** | **25,WE** |
| **18,WE** | **26,TH** |
| **19,TH** | **27,FR** |
| **20,FR** | 28,SA |
| 21,SA | 29,SU<br>DVENT BEGINS |
| 22,SU | **30,MO**<br>ST ANDREW'S DAY (SCOTLAND) |
| **23,MO** | |

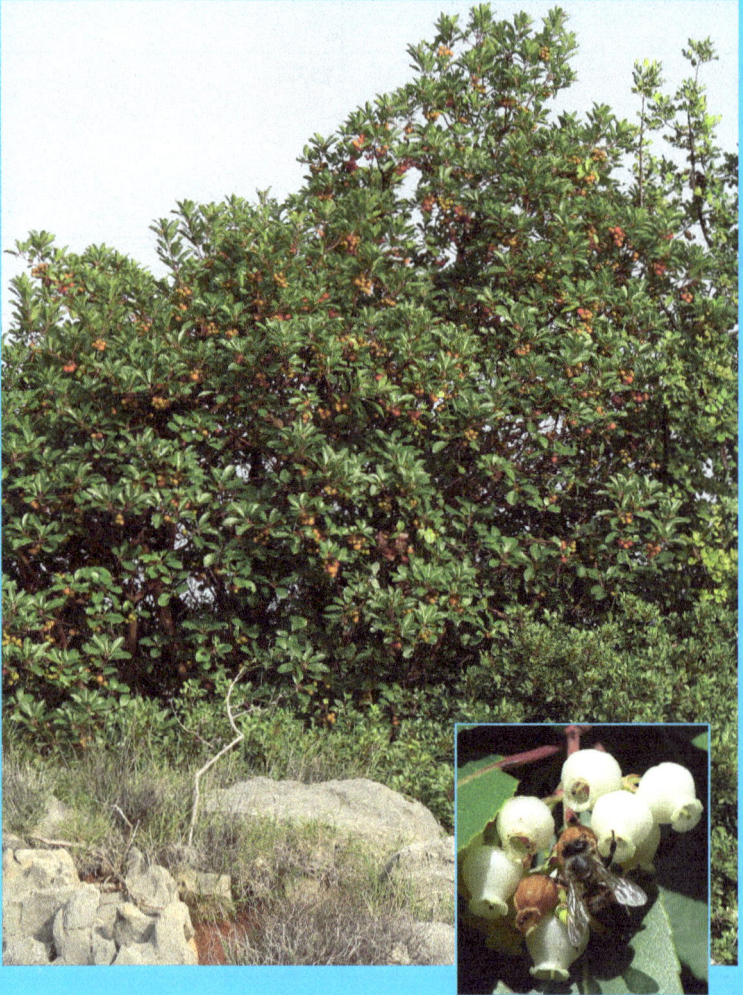

**Strawberry Tree, *Arbutus unedo***
This is one of my favourite trees in Greece. It flowers in mid-winter and looks wonderful
in the low light with its red strawberry-like fruits and its pure white bell-like flowers.
The bees seem to visit the flowers even when they look shrivelled up, so nectar continues
to be available for some time. The distinctive bitter honey is appreciated by many people who
dislike very sweet honey. The tree is hardy throughout most of the UK where the temperature
doesn't fall lower than -10C. Prefers a sheltered position in full sun.

| DAY | DECEMBER 2019 FORAGE | TEMP MIN | MAX | WIND DIR | B.S | CL'D | RAIN | 1 HIVE WEIGHT | 2 | 3 |
|---|---|---|---|---|---|---|---|---|---|---|
| 1 | | | | | | | | | | |
| 2 | | | | | | | | | | |
| 3 | | | | | | | | | | |
| 4 | | | | | | | | | | |
| 5 | | | | | | | | | | |
| 6 | | | | | | | | | | |
| 7 | | | | | | | | | | |
| 8 | | | | | | | | | | |
| 9 | | | | | | | | | | |
| 10 | | | | | | | | | | |
| 11 | | | | | | | | | | |
| 12 | | | | | | | | | | |
| 13 | | | | | | | | | | |
| 14 | | | | | | | | | | |
| 15 | | | | | | | | | | |
| 16 | | | | | | | | | | |
| 17 | | | | | | | | | | |
| 18 | | | | | | | | | | |
| 19 | | | | | | | | | | |
| 20 | | | | | | | | | | |
| 21 | | | | | | | | | | |
| 22 | | | | | | | | | | |
| 23 | | | | | | | | | | |
| 24 | | | | | | | | | | |
| 25 | | | | | | | | | | |
| 26 | | | | | | | | | | |
| 27 | | | | | | | | | | |
| 28 | | | | | | | | | | |
| 29 | | | | | | | | | | |
| 30 | | | | | | | | | | |
| 31 | | | | | | | | | | |

| DEC20 | 8,TU<br>FEAST OF THE IMMACULATE CONCEPTION |
|---|---|
| 1,TU | 9,WE |
| 2,WE | 10,TH |
| 3,TH | 11,FR |
| 4,FR | 12,SA |
| 5,SA | 13,SU |
| 6,SU<br>ST NICHOLAS | 14,MO |
| 7,MO<br>ST AMBROSE (PATRON SAINT OF BEEKEEPERS) | 15,TU |

| | |
|---|---|
| **16,WE** | **24,TH** |
| **17,TH** | **25,FR**<br>CHRISTMAS DAY |
| **18,FR** | 26,SA<br>BOXING DAY |
| 19,SA | 27,SU |
| 20,SU | **28,MO**<br>DEC BOXING DAY HOLIDAY |
| **21,MO** | **29, TU** |
| **22,TU** | **30, WE** |
| **23,WE** | **31,TH**<br>DEC NEW YEAR'S EVE |

| Hive/ Q NO. | Year Q Raised | Frames of Brood Autumn 2019 | Combs Covered | Honey Stored- Sugar fed Kg | Combs Covered Spring 2020 | Frames of Brood Spring 2020 | Spring Feeding Kg | Queens Reared | Nuclei |
|---|---|---|---|---|---|---|---|---|---|
| 1 | | | | | | | | | |
| 2 | | | | | | | | | |
| 3 | | | | | | | | | |
| 4 | | | | | | | | | |
| 5 | | | | | | | | | |
| 6 | | | | | | | | | |
| 7 | | | | | | | | | |
| 8 | | | | | | | | | |
| 9 | | | | | | | | | |
| 10 | | | | | | | | | |
| 11 | | | | | | | | | |
| 12 | | | | | | | | | |
| 13 | | | | | | | | | |
| 14 | | | | | | | | | |
| 15 | | | | | | | | | |
| 16 | | | | | | | | | |
| 17 | | | | | | | | | |
| 18 | | | | | | | | | |
| 19 | | | | | | | | | |
| 20 | | | | | | | | | |
| 21 | | | | | | | | | |
| 22 | | | | | | | | | |
| 23 | | | | | | | | | |
| 24 | | | | | | | | | |

# HONEYBEE COLONIES

| | | | | | | | | | |
|---|---|---|---|---|---|---|---|---|---|
| 1 | | | | | | | | | |
| 2 | | | | | | | | | |
| 3 | | | | | | | | | |
| 4 | | | | | | | | | |
| 5 | | | | | | | | | |
| 6 | | | | | | | | | |
| 7 | | | | | | | | | |
| 8 | | | | | | | | | |
| 9 | | | | | | | | | |
| 10 | | | | | | | | | |
| 11 | | | | | | | | | |
| 12 | | | | | | | | | |
| 13 | | | | | | | | | |
| 14 | | | | | | | | | |
| 15 | | | | | | | | | |
| 16 | | | | | | | | | |
| 17 | | | | | | | | | |
| 18 | | | | | | | | | |
| 19 | | | | | | | | | |
| 20 | | | | | | | | | |
| 21 | | | | | | | | | |
| 22 | | | | | | | | | |
| 23 | | | | | | | | | |
| 24 | | | | | | | | | |

# BEEEKEEPING RECORDS

| Number | items | Est. Value £ | P |
|---|---|---|---|
| | Stocks of Bees | | |
| | | | |
| | Empty Hives | | |
| | | | |
| | Combs - Deep<br>        - Shallow | | |
| | | | |
| | | | |
| | | | |
| | Frames | | |
| | Foundations | | |
| | Honey Extractor | | |
| | Honey Tanks | | |
| | Other items | | |
| | | | |
| | | | |
| | | | |
| | | | |
| | | | |
| | Honey Jars | | |
| | Honey | | |

## JANUARY 2021

| S | M | T | W | T | F | S |
|---|---|---|---|---|---|---|
|  |  |  |  |  | 1 | 2 |
| 3 | 4 | 5 | 6 | 7 | 8 | 9 |
| 10 | 11 | 12 | 13 | 14 | 15 | 16 |
| 17 | 18 | 19 | 20 | 21 | 22 | 23 |
| 24 | 25 | 26 | 27 | 28 | 29 | 30 |
| 31 |  |  |  |  |  |  |

## FEBRUARY 2021

| S | M | T | W | T | F | S |
|---|---|---|---|---|---|---|
|  | 1 | 2 | 3 | 4 | 5 | 6 |
| 7 | 8 | 9 | 10 | 11 | 12 | 13 |
| 14 | 15 | 16 | 17 | 18 | 19 | 20 |
| 21 | 22 | 23 | 24 | 25 | 26 | 27 |
| 28 |  |  |  |  |  |  |

## MARCH 2021

| S | M | T | W | T | F | S |
|---|---|---|---|---|---|---|
|  | 1 | 2 | 3 | 4 | 5 | 6 |
| 7 | 8 | 9 | 10 | 11 | 12 | 13 |
| 14 | 15 | 16 | 17 | 18 | 19 | 20 |
| 21 | 22 | 23 | 24 | 25 | 26 | 27 |
| 28 | 29 | 30 | 31 |  |  |  |

## APRIL 2021

| S | M | T | W | T | F | S |
|---|---|---|---|---|---|---|
|  |  |  |  | 1 | 2 | 3 |
| 4 | 5 | 6 | 7 | 8 | 9 | 10 |
| 11 | 12 | 13 | 14 | 15 | 16 | 17 |
| 18 | 19 | 20 | 21 | 22 | 23 | 24 |
| 25 | 26 | 27 | 28 | 29 | 30 |  |

## MAY 2021

| S | M | T | W | T | F | S |
|---|---|---|---|---|---|---|
|  |  |  |  |  |  | 1 |
| 2 | 3 | 4 | 5 | 6 | 7 | 8 |
| 9 | 10 | 11 | 12 | 13 | 14 | 15 |
| 16 | 17 | 18 | 19 | 20 | 21 | 22 |
| 23 | 24 | 25 | 26 | 27 | 28 | 29 |
| 30 | 31 |  |  |  |  |  |

## JUNE 2021

| S | M | T | W | T | F | S |
|---|---|---|---|---|---|---|
|  |  | 1 | 2 | 3 | 4 | 5 |
| 6 | 7 | 8 | 9 | 10 | 11 | 12 |
| 13 | 14 | 15 | 16 | 17 | 18 | 19 |
| 20 | 21 | 22 | 23 | 24 | 25 | 26 |
| 27 | 28 | 29 | 30 |  |  |  |

## JULY 2021

| S | M | T | W | T | F | S |
|---|---|---|---|---|---|---|
|  |  |  |  | 1 | 2 | 3 |
| 4 | 5 | 6 | 7 | 8 | 9 | 10 |
| 11 | 12 | 13 | 14 | 15 | 16 | 17 |
| 18 | 19 | 20 | 21 | 22 | 23 | 24 |
| 25 | 26 | 27 | 28 | 29 | 30 | 31 |

## AUGUST 2021

| S | M | T | W | T | F | S |
|---|---|---|---|---|---|---|
| 1 | 2 | 3 | 4 | 5 | 6 | 7 |
| 8 | 9 | 10 | 11 | 12 | 13 | 14 |
| 15 | 16 | 17 | 18 | 19 | 20 | 21 |
| 22 | 23 | 24 | 25 | 26 | 27 | 28 |
| 29 | 30 | 31 |  |  |  |  |

## SEPTEMBER 2021

| S | M | T | W | T | F | S |
|---|---|---|---|---|---|---|
|  |  |  | 1 | 2 | 3 | 4 |
| 5 | 6 | 7 | 8 | 9 | 10 | 11 |
| 12 | 13 | 14 | 15 | 16 | 17 | 18 |
| 19 | 20 | 21 | 22 | 23 | 24 | 25 |
| 26 | 27 | 28 | 29 | 30 |  |  |

## OCTOBER 2021

| S | M | T | W | T | F | S |
|---|---|---|---|---|---|---|
|  |  |  |  |  | 1 | 2 |
| 3 | 4 | 5 | 6 | 7 | 8 | 9 |
| 10 | 11 | 12 | 13 | 14 | 15 | 16 |
| 17 | 18 | 19 | 20 | 21 | 22 | 23 |
| 24 | 25 | 26 | 27 | 28 | 29 | 30 |
| 31 |  |  |  |  |  |  |

## NOVEMBER 2021

| S | M | T | W | T | F | S |
|---|---|---|---|---|---|---|
|  | 1 | 2 | 3 | 4 | 5 | 6 |
| 7 | 8 | 9 | 10 | 11 | 12 | 13 |
| 14 | 15 | 16 | 17 | 18 | 19 | 20 |
| 21 | 22 | 23 | 24 | 25 | 26 | 27 |
| 28 | 29 | 30 |  |  |  |  |

## DECEMBER 2021

| S | M | T | W | T | F | S |
|---|---|---|---|---|---|---|
|  |  |  | 1 | 2 | 3 | 4 |
| 5 | 6 | 7 | 8 | 9 | 10 | 11 |
| 12 | 13 | 14 | 15 | 16 | 17 | 18 |
| 19 | 20 | 21 | 22 | 23 | 24 | 25 |
| 26 | 27 | 28 | 29 | 30 | 31 |  |

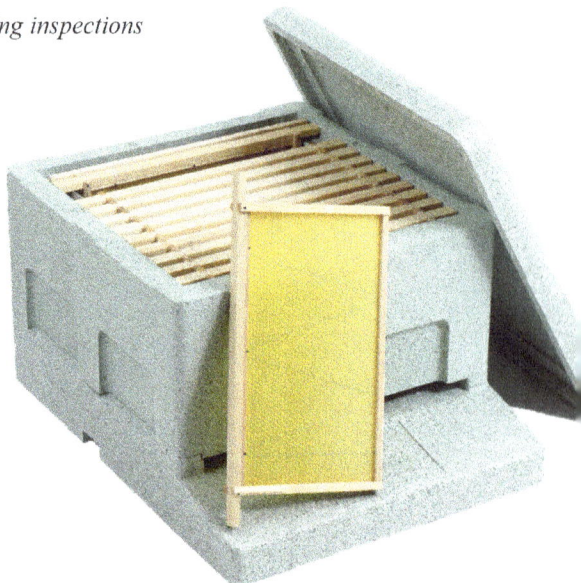

# DIRECTORY 20

## DIRECTORY, Associations and Services

Every effort is made to keep entries up to date but the publishers cannot be held responsible for errors or omissions.
Associations and all other groups listed have been requested (August 2019) to supply updated entries.
Readers who are aware of inaccuracies are asked to send updates to jerry@northernbeebooks.co.uk

# BEEKEEPING ASSOCIATIONS
## Bee Educated e-learning for Beekeepers

http://www.beeeducated.co.uk/
ian@BeeEducated.co.uk

### EDU
**e-Learning for Beekeepers**
*A module website set-up specifically for beekeeping tutors and their students*
http://www.beeeducated.co.uk/
mail: st@zbee.com

### BDI
**Bee Disease Insurance Ltd**
http://www.beediseasesinsurance.co.uk/home
donald.robertson-adams@beediseasesinsurance.co.uk

### BEES
**Beekeeping Editors' Exchange Scheme**
*Helping editors help themselves*
editors-owner@ebees.org.uk

### BA
**Bees Abroad**
*Relieving poverty through beekeeping*
http://beesabroad.org.uk/
info@beesabroad.org.uk
www.facebook.com/beesabroad/

### BFA
**Bee Farmers Association**
*The voice of professional beekeeping*
http://beefarmers.co.uk/
gensec@beefarmers.co.uk
Publication: Bee Farmer Journal
www.facebook.com/beefarmersassociation

## B for D

**Bees for Development**
*The specialist international beekeeping organisation*
http://www.beesfordevelopment.org/
info@beesfordevelopment.org
www.facebook.com/beesfordevelopment/
Publication: Bees for Development Journal

## BBKA

**British Beekeepers Association**
https://www.bbka.org.uk/
www.facebook.com/groups/BBKA.info/
Publication: BBKA News

## BIBBA

**British Improvement and Bee Breeders' Association**
*For the conservation, restoration, study, selection and improvement of native or near native honeybees of Britain and Ireland*
https://bibba.com/
membership@bibba.com
Publication: Bee Improvement Magazine
www.facebook.com/beeimprovement

## CABK

**Central Association of Beekeepers**
https://www.cabk.org.uk/
pamhunter@burnthouse.org.uk
Publications: Selected lectures

## CBDBBRT

**C.B. Dennis British Beekeepers Trust**
https://sites.google.com/site/cbdennistrust/
cbdennisbeetrust@gmail.com

## CONBA

**Council of National Beekeeping Association of the UK**
*Its purpose is to represent the interests of beekeepers with local, national and international authorities*
http://www.conba.org.uk/
mail: philmcanespie@btinternet.com

## DARG

**Devon Apicultural Research Group**

*Combining practical beekeeping with leading edge apicultural and environmental science*

http://dargbees.org.uk/

## ECT

**The Eva Crane Trust**

*To advance the understanding of bees and beekeeping*

https://www.evacranetrust.org/

mail@evacranetrust.org

## FIBKA

**Federation of Irish Beekeepers' Association**

https://irishbeekeeping.ie/

secretary@irishbeekeeping.ie

www.facebook.com/fibka/

Publication: An Beachaire

## IBRA

**International Bee Research Association**

*Promotes the value of bees by providing information on bee science and beekeeping worldwide*

http://www.ibrabee.org.uk/

mail@ibra.org.uk

www.facebook.com/IBRAssociation/

Publications: Bee World, Journal of Apicultural Research

## INIB

**Institute of Northern Ireland Beekeepers**

http://www.inibeekeepers.com/

membershipsecretary@inibeekeepers.com

## LASI

**Laboratory of Apiculture and Social Insects**

*The applied research is aimed at helping the honey bee and beekeepers, whilst the basic research studies how insect societies function*

http://www.sussex.ac.uk/lasi/

www.facebook.com/

## NBKT
**Natural Beekeeping Trust**
*We encourage attention to the real nature of bees, their nesting preferences, their forage needs and their all-encompassing purpose.*
https://www.naturalbeekeepingtrust.org/
www.facebook.com/naturalbeekeepingtrust/

## NDB
**National Diploma in Beekeeping**
*The NDB exists to meet a need for a beekeeping qualification above the level of the Certificates awarded by the United Kingdom National Beekeeping Associations*
https://national-diploma-beekeeping.org/

## NHS
**National Honey Show**
*Promoting the highest quality honey and wax products with international classes,lecture convention, workshops and beekeeping equipment trade show*
http://www.honeyshow.co.uk/

## NIHBS
**Native Irish Honey Bee Society**
*To support the various strains of Native Irish Honey Bee (Apis mellifera mellifera) throughout the country. It is a cross border organisation and is open to all. It consists of members and representatives from all corners of the island of Ireland*
http://nihbs.org/
secretary@nihbs.org
www.facebook.com/native-irish-honey

## SBA
**Scottish Beekeepers' Association**
*The organisation's purposes are to support honey bees and beekeepers, to improve the standard of beekeeping, and to promote honey bee products in Scotland*
https://scottishbeekeepers.org.uk/
secretary@scottishbeekeepers.org.uk
Publication: The Scottish Beekeeper

## UBKA
**Ulster Beekeepers' Association**
https://www.ubka.org/
ubkasecretary@gmail.com
www.facebook.com/Ulsterbees/

## WBKA
**Welsh Beekeepers' Association**
*We represent Welsh beekeepers nationally within Wales and the UK and internationaly.*
http://www.wbka.com/
secretary@wbka.com
Publication: Welsh Beekeeper

## HONEY BEE HEALTH
**England and Wales**

## NBU
## National Bee Unit
http://www.nationalbeeunit.com/
nbu@apha.gsi.gov.uk

## Northern Ireland
## DARD
**Department of Agriculture Northern Ireland**
https://www.daera-ni.gov.uk/

## Scotland
## SG-AFRC
**The Scottish Government Rural Payments and Inspections Directorate**
bees_mailbox@gov.scot

# FACEBOOK BEEKEEPING GROUPS

Beekeeping for Beginners
Beekeeping Basics
Friendly Beekeepers
Beekeeping Hacks
Women in Beekeeping
The Beekeepers Bulletin
Poly hive beekeeper
Commercial Beekeepers
HAPPY BEEKEEPERS
UK Beekeepers
Cold Climate Beekeeping
Stewart's Beekeeping Basics
Newbie Beekeeping
Warre Beekeeping
Scientific Beekeeping
Backyard Beekeeping
British Beekeepers
Scottish Beekeepers
Beekeeping
Sideliner Beekeepers
Loui's Mountain Beekeeping
Beekeepers of Ireland
Beekeeping for beginners to experienced
London Beekeepers Association
Beekeeping Tools, Supplies & Hardware
Beekeeping with the warré hive
Beekeeping Classifieds
Florida Beekeepers
Northern Beekeeping
Irish Beekeepers' Association
Beekeeping Wood Shop
Buy/Sell Beekeeping Equipment (Serious Buyers)
Beekeeper Builders Corner
Top Bar Hive Beekeepers
Beekeeping Victoria (Australia)
Beekeeping for Beginners UK
Beekeeping Top Tips UK
Beekeepers buying or selling UK
Beekeeping in France
"OTS" Beekeeping
Lincolnshire Beekeeping Association- Sleaford District

Backyard beekeeping NZ
Beekeeping Science
Too Big to Quit Now - Sideliner Beekeepers
Beekeeping
Backyard Beekeeping Australia
Beekeeping for all
Beekeeping Questions UK
Treatment-free beekeepers
Beekeepers Trading Post
Better Beekeeping
Beekeeping Techniques
Beekeeping Apimarket UK
beekeeping Warm and cold Large and small
Beekeeping Classifieds Victoria Australia
Jersey Beekeepers' Association
Black bee beekeepers
Beekeeping Equipment For Sale Australia
Northamptonshire Treatment Free BeeKeeping
Beekeeping
Modern BeeKeepers
Beekeepers Group :D
Beekeeping Petra
amateur backyard beekeeping
Beekeeping Supplies eBay/Amazon
World Beekeepers Collective
Beekeeping for all
Urban Beekeeper
NWA Beekeeping Alliance
Beekeeping Basics
Successful Beekeeping
Beekeeping Questions
Irish Beekeeping
EN-Beekeepers
Ohio beekeepers
Beekeeping Qld Australia
BEEKEEPERS WORLD CLUB
Natural and Balanced Beekeeping DIscussion Group
Buckfast Beekeepers Group
Beekeeping
Commercial Beekeepers Trucks And Equipment For Sale Trade Etc
Louth Beekeepers
Preservation Beekeeping Community Page
Beekeepers With Power!!

APICULTRICES - WOMEN IN BEEKEEPING
Beekeepers Kenya
UK Beekeepers: Buy & Sell
BeekeepersOfCT
Beekeeping Photography
Beekeepers of Ireland BUY/SELL/SWAP/FREE
COBA Central Oklahoma Beekeepers Association

## Suppliers of Beekeeping Equipment, etc.
www.facebook.com/E.H.Thorne/
www.facebook.com/BJSherriff/
www.facebook.com/BeeEquipmentLtd/
www.facebook.com/cornishhoney.co.uk/
www.facebook.com/quincehoneyfarm/
www.facebook.com/NaturalApiary/
www.facebook.com/BBwear/
www.facebook.com/paynesbeefarm/
www.facebook.com/maisemoreapiaries/
www.facebook.com/nationalbeesupplies/
www.facebook.com/BSHoneyBees/
www.facebook.com/Beckysbees/

# ALPHABETICAL QUIZ ANSWERS

A: *Malpighamoeba mellificae*

B:  Burtt

C:  Certan

D:  Dzierzon, Johann

E:  *Euodia* (formerly Evodia)

F:  Figwort

G:  Glossa

H:  Hypopharyngeal

I:  *Impatiens glandulifera*, Himalayan balsam

J:  Janska, Anton

K:  Killion

L:  *Limnanthes douglasii*, Poached egg plant

M:  UK city which featured bee sculptures in 2018

N:  Nelson, J

O:  Orlando

P:  Pettigrew

Q:  Quinby, Moses

R:  R: Root

S:  Sacbrood

T:  Tracheal

U:  Uncapping knife

V:  Ventriculous

W: Weiss

X:  Xylene

Y:  Yorkshire spacer

Z:  Zander

www.ingramcontent.com/pod-product-compliance
Lightning Source LLC
Chambersburg PA
CBHW071447200326
41519CB00019B/5645